## ONLY ISAAC ASIMOV COULD BRING TO-GETHER SUCH DIVERSE BRANCHES OF SCIENCE AS . . .

Archaeology • Zoology • Geology • Chemistry • Nuclear Physics • Astronomy • Cosmology • Biology •

. . . and much more!

## ONLY ISAAC ASIMOV COULD SO LUCIDLY AND ENTERTAININGLY EXPLAIN THE ORIGINS OF SO MANY THINGS . . .

Amphibians • Marsupials • Bacteria • DNA • Energy • Light • Stars • Galaxies • Farming • Toolmaking • Writing • Human Flight • Homo sapiens • Homo habilis • Homo erectus •

. . . and much more!

ISAAC ASIMOV is recognized worldwide for his ability to illuminate the most complex concepts and make science enjoyable and stimulating to readers of all ages. He is the author of 375 books, including THE HISTORY OF PHYSICS, THE COLLAPSING UNIVERSE, and ASIMOV'S GUIDE TO HALLEY'S COMET.

# *BEGINNINGS*

# The Story of Origins— of Mankind, Life, the Earth, the Universe

# ISAAC ASIMOV

**B**
BERKLEY BOOKS, NEW YORK

*The author wishes to acknowledge the assistance of Sandra Kitt in preparation of the diagrams.*

This Berkley book contains the complete
text of the original hardcover edition.
It has been completely reset in a typeface
designed for easy reading and was printed
from new film.

BEGINNINGS

A Berkley Book / published by arrangement with
Walker and Company

PRINTING HISTORY
Walker and Company edition published 1987
Berkley edition / May 1989

ISBN: 0-425-11586-0

A BERKLEY BOOK® TM 757,375
Berkley Books are published by The Berkley Publishing Group,
200 Madison Avenue, New York, NY 10016.
The name "BERKLEY" and the "B" logo
are trademarks belonging to Berkley Publishing Corporation.

PRINTED IN THE UNITED STATES OF AMERICA

10  9  8  7  6

*To the Smiley family*
*For making us happy at Mohonk*

# CONTENTS

Introduction   ix

What are the Beginnings of . . .

1. Human Flight   1
2. History   7
3. Civilization   20
4. Modern Man   37
5. Homo Sapiens   42
6. Hominids   57
7. Primates   69
8. Mammals   76
9. Animal Flight   87
10. Reptiles   97
11. Land Life   114
12. Chordates   126
13. Continents   136
14. Earth   157
15. Fossils   167
16. Multicellular Organisms   172
17. Eukaryotes   184
18. Prokaryotes   191
19. Viruses   196
20. Ocean and Atmosphere   202
21. Life   219
22. Moon   239
23. Solar System   246
24. Universe   255

Index   275

# INTRODUCTION

In writing a book about beginnings, I start with one enormous advantage. All the governments of the world agree on the manner of measuring time.

The years are numbered consecutively, so that since I am writing this sentence in 1987, I know that last year was 1986 and that next year will be 1988, and that no one will disagree with this.

Each year is divided into twelve months and a given month may have anything from twenty-eight to thirty-one days. This is an unnecessary irregularity but it is one upon which the whole world agrees, and if I say this day is February 2, 1987, in New York City, everyone will agree (though in parts of the world, at this moment, February 3 is already considered to have been begun). Again, we are all agreed that the year begins on January 1.

This is not to say there are not special calendars used by various religious faiths or by different nations that follow older traditional ways of treating time, but these are all local and special. While they add a flavor of variety and interest to human affairs, they do not confuse matters. In all formal dealings, the International Calendar is used. It is called the Gregorian Calendar because its final details were officially put in place by Pope Gregory XIII in 1582.

This was not always so. It is only in relatively recent times that the matter of timekeeping was accepted and made practically universal, but at least it enables us to look backward from a firmly established present.

In this book, I'm going to take up beginnings of various kinds, starting with something relatively mundane and everyday, and moving on steadily to matters that are more sweeping and general until we end, finally, by considering the possible time and events of the beginning of the Universe itself.

Each chapter that follows, will be devoted to the beginning of something and that something will be the name of the chapter. We will start with a specific human technology that is fully documented and that, therefore, should offer us no problem.

# BEGINNINGS

# 1

## HUMAN FLIGHT

In a large city such as New York, Chicago, or Los Angeles, one can look up at any hour of the day or night and see one or more airplanes (or, at night, their lights) moving across the sky. It is so common a sight that no one pays any attention.

Yet when I was a small boy in the 1920s, the sight of an airplane in the sky over New York was so unusual that people rushed out of their houses to watch it and marvel. Airplanes must have begun flying, then, not very long before the 1920s. When, actually, did they start? When did human flight begin?

The answer would seem to be simple. On December 17, 1903, an American inventor, Orville Wright (1871–1948), made the first airplane flight in history at Kitty Hawk, North Carolina. He had built an airplane together with his brother, Wilbur Wright (1867–1912). That first airplane flew only 850 feet, barely skimming the ground. It was in the air for less than a minute and went slowly enough so that Wilbur could run alongside. That was the first successful flight in an airplane, and it might be said to represent the beginning of human flight.

Does that close the case? Can we now drop the matter of human flight and pass on to a new subject?

No, for the matter is not quite that simple. The Wright brothers were not working in a vacuum. Other people were investigating the matter, too.

The American astronomer Samuel Pierpont Langley (1834–1906) began experimenting with airplanes in 1896, and before the Wright brothers' flight he had made three attempts to fly his planes. The third time he nearly made it, but not quite. In 1914, his third plane was fitted with a more powerful engine and was successfully flown, but Langley was dead by then.

Well, then, did airplane flight begin, perhaps, with Langley's near-success?

We might answer the question in this way. Langley is certainly an honored part of the history of human flight. So were still earlier investigators who labored to build flying machines or who worked out the scientific rules that made such machines possible. After all, about 1500, the Italian engineer and artist Leonardo da Vinci (1452–1519) made interesting drawings of flying machines, based on an intelligent consideration of mechanics. And the ancient Greeks, two thousand years earlier still, invented fanciful tales of constructing feathered wings that enabled men to fly. However, the true beginning would be the first *successful* flight that was followed up by other successful flights.

And yet, having said all that, we must admit that Orville Wright was not the first human being to fly successfully. He was the first to fly a vehicle that was heavier than air: one that flew despite the fact that it would not float in the air. But what about vehicles that *did* float in the air?

On July 2, 1900, the German inventor Ferdinand von Zeppelin (1838–1917) successfully launched the first flight, in which a gondola capable of holding human beings was suspended beneath a hydrogen-filled, cigar-shaped bag that could float in the air. Such a device was a *dirigible balloon,* or simply a *dirigible,* from the Latin word meaning "to direct." Since such a device was outfitted with an internal combustion motor and a propeller, its motion could be directed in any direction, even against the wind. Such devices were also called *zeppelins* after the inventor, and *airships* for obvious reasons.

Additional dirigibles were built, and they were used for commercial airflights before airplanes were. In the 1920s and early 1930s, they seemed to represent the direction in which human flight was going. Why is it, then, that the start of human flight is always given as Wright's airplane flight in 1903, rather than as von Zeppelin's flight in 1900?

The answer is that, in the end, dirigibles lost the race. The hydrogen-filled bag was too susceptible to fire, as was shown when the *Hindenburg*, the largest dirigible ever built, suddenly burst into flame as it was docking at Lakehurst, New Jersey, on May 6, 1937. Even dirigibles making use of nonflammable helium in their bags were too vulnerable to storms. Dirigibles passed from the scene before World War II, therefore, just as airplanes were becoming ever larger and faster.

Dirigibles, as the unsuccessful competitor in human flight, tend therefore to be forgotten, and the beginning of flight is always marked as Orville Wright's airplane flight.

But let us go farther back in time.

In 1852, forty-eight years before Zeppelin, a French engineer, Henri Giffard (1825–1882), placed a steam engine in a gondola under a sausage-shaped balloon and had it turn a propeller so that it could move in any desired direction at 6 miles per hour.

Should that, then, be considered the first dirigible flight? No, because nothing ever came of Giffard's device. It was what we might call a "laboratory demonstration" that wasn't really practical. It could be done, but it wasn't worthwhile doing it. That is why one should define a true beginning not only as an event that was successful, but one that was followed by *other* events of the sort, one that "caught on."

Why did Zeppelin's invention catch on, while Giffard's didn't? For one thing, Zeppelin did not work with a bare bag of hydrogen but encased the bag in thin aluminum shells, making it mechanically much stronger and making it possible to streamline it more efficiently so that it could move more quickly. Zeppelin also used an internal-combustion engine rather than a steam engine, and the former was more efficient. Though, to be sure, neither aluminum nor internal-combustion engines were available to Giffard, so he cannot be faulted too strenuously for not taking advantage of such things.

Yet even so, disregarding Giffard, it remains true that human beings were flying, successfully and practically, before the Wrights and before von Zeppelin in devices that were neither airplanes nor dirigibles.

Both airplanes and dirigibles, after all, are powered devices that can force their way against the wind, but what about *unpowered* devices that get their only motive power from the wind?

Airplanes without engines are called *gliders*. When launched from a hilltop or cliff, gliders can, if properly designed, coast for considerable distances, especially if they take advantage of updrafts. The Wright brothers flew gliders many times before they flew an airplane. Their first airplane was, in fact, very little more than an improved glider outfitted with an internal-combustion engine.

Again, dirigibles without engines are called *balloons,* and these, floating in air, could drift with the wind and carry human beings for considerable distances long before powered flight was invented.

The English engineer George Cayley (1773–1857) was the first to study scientifically the conditions under which air might keep an artificial device aloft. He thus founded the science of aerodynamics. He was the first to realize that what was needed were fixed wings (like the flaps of a flying squirrel), rather than movable wings (like those of birds). He worked out the basic shape that airplanes would eventually have—wings, tail, streamlined fuselage, and rudder—and realized that if it could be made light enough, the wind would carry the device through the air for extended flights. He also realized that it needed an engine and propeller to be able to go against the wind, but he knew that no engine then existing would be light enough and powerful enough.

In any case, in 1853, he built the first glider capable of carrying a man through the air. He was sixty years old at the time and did not feel up to chancing an actual flight (or perhaps he valued his neck too much). In those days, however, servants were expected to obey orders. Cayley therefore ordered his coachman (against the poor man's vehement objections) to take the first glider flight. The coachman did, and survived.

This was one year after Giffard's first powered balloon flight, but Cayley's unpowered glider came to something. Other and still better gliders were built, and by the end of the nineteenth century, gliding had become a popular sport among the young and adventurous. The most famous gliding devotee of the time was a German engineer, Otto Lilienthal (1848–1896), who died of injuries when his glider finally crashed.

However, before Cayley's unpowered airplane, there were unpowered balloons.

The first successful balloons were built in 1783 by two brothers, Joseph Michel Montgolfier (1740–1810) and Jacques Etienne Montgolfier (1745–1799). The first balloon (filled with hot air) flew on June 5, 1783, but it was not until November 20, 1783, that a large enough balloon was built that could carry a human being—two human beings, in fact. One was a young physicist, Jean Francois Pilatre de Rozier (1756–1785) and the other was one Marquis d'Arlande. They were the first human beings to fly through the air in a human device, the first "aeronauts," no less than 120 years before the Wrights.

On January 7, 1785, Pilatre de Rozier was carried, with two others, across the English Channel by balloon. When he tried to return by balloon on June 15, the fire used to heat the air in the balloon (to keep it lighter than ordinary air) set the balloon's fabric aflame, and he fell to his death from the height of nearly a mile. The first aeronaut was thus the first to die in an aeronautical disaster.

As you see from this account, then, deciding on the

beginning of even a quite modern phenomenon isn't so easy, not even when you have all the dates. You have to be clear as to what it is you are tracing the beginning of—heavier-than-air powered machines, lighter-than-air powered machines, or unpowered machines. You must decide whether to include unsuccessful attempts, or successful ones that lead to no consequences.

Another point we might make is that beginnings can be a little fuzzy because changes almost invariably come through a process of evolution—that is, an accumulation of small changes, sometimes so small that you can't specify the point at which you can say, "Here is the beginning."

This is true of almost anything, and it becomes obviously true as whatever it is you are tracing back to a beginning becomes broader. For instance, suppose it was not powered flight you were tracing back to its beginning, but history itself. When does the careful account of the battles and the struggles, the problems and the solutions, the malignant evil and the labored good, that marks the long story of humanity *begin?*

Every American schoolchild can think back to 1776, when the American colonies declared their independence, and even to 1492, when Christopher Columbus (1451–1506) discovered the New World. But surely that is not as far back as we can go. Columbus's discovery was not quite five hundred years ago, and history extends into far earlier times, during which Europeans didn't dream that the American continents existed.

Let's therefore reach back into time in search of some moment when history began.

# 2

## HISTORY

Western Europe, at the time of Columbus's voyage was just entering "modern times." In fact, 1492, precisely because Columbus's epoch-making discovery took place in that year, is often considered the actual beginning of modern times. Of course, like beginnings generally, this is largely a matter of definition. Quite good arguments can be presented for beginning modern times as early as 1453 (the fall of Constantinople to the Turks) or as late as 1517 (the beginning of the Protestant Reformation). However, we shall accept 1492 without further discussion.

Written documentation is excellent in modern times. For one thing, so little time has passed that there has not been much chance of crucial documentation being permanently lost or destroyed.

For another thing, the German inventor Johannes Gutenberg (1398–1468) invented printing with movable type around 1450, and with that it became possible to multiply records of all types to such an extent that permanent loss and destruction was all but impossible.

Before modern times, however, there was a thousand-year stretch of time in western Europe that is usually referred to as the medieval period or the Middle Ages, because it comes between modern times and ancient times. The Middle Ages, particularly the first half, is rather arid in documentation. For one thing there has been more time to bring about losses and more vicissitudes to occasion them, especially in the absence of printing. Then, too, it was an "age of faith," in which matters relating to religion were considered far more important than matters relating to the world, so that fewer and poorer records were kept.

Nevertheless, though history is a little fuzzy in that thousand-year period, we have enough to outline events fairly well.

Modern Spain, for instance, was formed, in more or less its present form, only toward the close of the Middle Ages. Earlier, it had existed as a group of small Christian kingdoms in the northern part of the peninsula because of the shattering blow of an Islamic invasion from Africa early in the period. Slowly, the Christian lands grew at the expense of the Islamic (Moorish) lands in the south and coalesced. By the 1400s, there were three Christian kingdoms in the peninsula: Portugal to the west, Aragon to the east, and Castile (the largest) in the middle.

In 1469, Isabella (1451–1504), the heiress of Castile, married Ferdinand (1452–1516), the heir of Aragon, and in 1479, when each had succeeded to the throne, the two kingdoms were united and remained so. In 1492, just before Columbus's voyage, the united kingdom of Spain conquered the last of the Moorish regions in the south, and modern Spain was born.

England, in its modern form, is older. William, duke of Normandy (1027–1087), invaded England, defeated the English at the Battle of Hastings on October 14, 1066, and established a strong monarchy there. Queen Elizabeth II, who rules the land now, can trace her descent from William, so that the line has now endured for 921 years.

France traces its present-day form even further back, to the accession to the throne of Hugh Capet (940–996) in

987 (exactly a thousand years ago as I write this). The last of his descendants, Louis Philippe I, left the throne in 1848, so that the line persisted for 861 years.

Germany has had a very checkered history, during most of which it has consisted of fragments that were as likely to quarrel with each other as with non-German enemies. During the Middle Ages, however, they made up the core of a political structure called the Holy Roman Empire, and this was sometimes strong.

The Holy Roman Empire came into being when Charlemagne (742–814), the ruler of the Frankish kingdom, which then controlled western Europe, was crowned emperor in Rome on December 25, 800, by Pope Leo III (750–816).

Charlemagne, incidentally, was the ruler who ordered that the years be numbered according to the present system. The usage, which I praised in my introduction, was established in his broad dominions and, eventually, over the whole world. At this point, then, let me digress a moment and explain how the system works and why the year in which I write this is 1987 and not some other number.

In ancient times it was customary sometimes to identify a year by naming it for some notable event that took place in it. It might be called "the year of the great blizzard," for instance. The writer P. G. Wodehouse parodies this by frequently referring to a time as "the year such-and-such-a-horse won the Derby."

Naturally, except for the people who lived through the period and remember the event, such an identification is useless.

A somewhat more regular system is to identify the year by the reigning executive, usually a king. One might say "in the third year of the reign of Jehoshaphat," or "in the twenty-second year of the reign of Manasseh." Years are so identified in the Bible, which makes it difficult to convert biblical chronology into the ordinary system.

Obviously, the logical thing to do is to pick an event of particular importance and to number all the years on and on from that event, making no new start at any time.

The years as we number them today are just like that. They start at a particular event and are numbered on indefinitely.

Many people, however, don't realize that the year 1 merely memorializes some event but think that it really represents a beginning. People sometimes say, "Ever since the year one," meaning "As long as things have existed." I've even heard people refer casually to the Earth as being not quite two thousand years old.

If we did start counting from a year 1, a sensible rule would be to set it so far back in time that we are likely never to have occasion to worry about years that are still earlier. To see an example of this, let us move backward into ancient times.

In the most recent portion of ancient times, the Mediterranean shores (southern Europe, westernmost Asia, and northern Africa) were under the control of the Roman Empire, which ruled from Rome, Italy. The last Roman Emperor was dethroned in Italy in 476, and this is sometimes taken as the end of ancient times and the beginning of the Middle Ages.

Marcus Terentius Varro was a Roman who lived before the Roman Empire was established and while Rome was still ruled by elected consuls and by a Senate, so that it was called the Roman Republic.

Varro decided that it would make sense to start numbering from the year in which the city of Rome was founded. Since Romans rarely had occasion to talk about events before that founding, by using this system they would always have positive numbers to deal with and would almost never have to face the problem of numbering years earlier than 1.

Varro studied the Roman histories of the time and calculated the year in which the city of Rome must have been founded. He counted the lists of consuls who were said to have ruled the city, and the number of years that each of the legendary kings had ruled Rome in its very early history. Finally, he came up with a year for the founding of the city, called it 1, and numbered all succeeding years from that. This system of counting the years is called the "Roman Era" or the "Era of Varro."

When the Rome writers numbered the years in this fashion, they commonly added the initials A.U.C., which stood for *Anno Urbis Conditae,* meaning "the year of the founding of the city." Thus, Varro was born in 637 A.U.C. and died in 726 A.U.C. at the age of eighty-nine. As for Charlemagne, he was crowned in the year 1553 A.U.C.

In Christian times, however, there were those who didn't think that the founding of the city of Rome (which had been pagan for the first thousand years of its existence) was the proper mark from which to count the years. They felt, rather, that the birth of Jesus was the central event of history and that the year in which he was born should be the reference point of numbering.

The trouble was, though, that the year of Jesus' birth was uncertain. The Bible doesn't give the years according to the Roman era. It does give some guidance, however, and about 525 a monk named Dionysius Exiguus attempted to calculate the year of Jesus' birth.

Thus, in the Gospel of St. Luke, Jesus' birth is said to come at a time when the Emperor Augustus ordered that the population of the empire be taxed. It goes on to say, "And this taxing was first made when Cyrenius was governor of Syria."

Cyrenius (more accurately, Quirinius) was indeed in charge of Roman military affairs in Syria and Judea on two different occasions during the reign of Augustus. Augustus reigned 726–767 A.U.C. and Quirinius held the office 747–749 A.U.C. and again 759–762 A.U.C.

In the Gospel of St. Matthew, it states that Herod was ruling over Judea (as a Roman puppet, of course) at the time of Jesus' birth, and Herod ruled 716–749 A.U.C. The only years in which all three were in power were 747–749 A.U.C., so Jesus had to be born in that period if the biblical statements are correct.

Dionysius Exiguus, however, finally arrived at the figure 753 A.U.C. for the birth year of Jesus, and that was accepted by the Christian world. The fact that he had made a mistake of at least four years was not recognized until so many people had used and were using his system that it was impossible to change it.

If we assume Jesus was born on December 25, 753 A.U.C., then 754 A.U.C. is the year 1, 755 A.U.C. is 2, and so on, and we will eventually reach 1776 (2529 A.U.C. = 1776 + 753) as the year of the Declaration of Independence. To indicate that we are marking off the years from Jesus' birth, we say A.D. 1776, where A.D. stands for *Anno Domini*, which is Latin for "the year of the Lord."

The system can be called the Christian Era or the Dionysian Era. Some people who are not Christian prefer to call it the Common Era and to use the appropriate initials so that they might speak of 1776 C.E. Actually, however, the system is so common and so taken for granted that one hardly ever sees initials used with it—A.D. 1776 is merely 1776.

Actually, the Christian era has a grave flaw. The year 1 is uncomfortably late in history. Julius Caesar and everything before him is before the year 1. It is necessary to start counting backward. Thus, since Julius Caesar was assassinated forty-four years before A.D. 1, he was assassinated in 44 B.C., where the initials stand for *Before Christ*. As for the foundation of the city of Rome, that took place 753 years before A.D. 1 and, therefore, in 753 B.C. (Those non-Christians who don't want to memorialize Jesus use the initials B.C.E., which stands for *Before the Common Era*.)

A minor, but irritating, flaw in this system is that no provision was made for a year 0 dividing the A.D. from the B.C. If there had been a year 0, the first decade would have been 0 to A.D. 9 inclusive, and A.D. 10 would have started a new decade. Every decade would start on January 1 of a year ending with a 0; every century on January 1 of a year ending in 00, and every millennium on a year ending in 000.

Because there is no year 0, however, the first decade is A.D. 1 to A.D. 10 inclusive, and it is A.D. 11 that starts the second decade. The decades, centuries, and millennia all start on January 1 of years that end in 1, 01, and 001 respectively.

Thus, under present conditions, A.D. 2000 will be the last year of the second millennium, and the third will begin on January 1, 2001. We can be sure, however, that the entire world will celebrate the start of a new millennium on January 1, 2000, and that no amount of explanation to the effect that the celebration is exactly one year premature will help.

Then, too, since Jesus cannot have been born later than 749 A.U.C. if the statements in Matthew and Luke are correct, he can't have been born later than 4 B.C., that is, four years before his own birth. You will even find 4 B.C. given for Jesus' birth year in many editions of the Bible. This would surely create a laugh, if we were allowed to laugh at such a thing.

Histories written during the period of Roman dominance could carry us far back into ancient times. Of course, everything was handwritten then, so there were few copies of individual books and many have been lost altogether.

Even so, what has survived takes us back with considerable reliability to 390 B.C. (363 A.U.C.), at which time a band of Gauls (Celtic barbarians who were invading Italy) took and sacked Rome, which was then a small city that headed a confederation of even smaller neighboring cities.

The Roman records were pretty much destroyed by this barbarian incursion, so that references to earlier events in Roman history may be in part distorted and in part altogether legendary and fictional. (This is not surprising. There are events in early American history that every schoolchild, and almost every adult, believes, that are probably fiction. The story of George Washington and the cherry tree is certainly fiction, and the story of John Smith being saved by Pocahontas is very likely to be so).

Allowing for this, we have 509 B.C. (244 A.U.C.) as the traditional date of the founding of the Roman Republic. The seven kings who ruled Rome in the first two and a half centuries of its existence came to the end of the line when the seventh, Lucius Tarquinius Superbus, was overthrown and exiled. And, of course, 753 B.C. (1 A.U.C.) is the traditional date of the founding of Rome.

But does history extend back beyond the founding of Rome, too?

Different cities often have traditional dates for their founding, but there is a good chance that these dates are shoved back in time and made earlier than is so, simply because cities like to appear older and more venerable than they really are. It's a matter of prestige—and this may well be true of Rome itself.

Thus, the city of Carthage, the great rival of Rome in the third century B.C., gave out the traditional date of its founding as 814 B.C. This would make it sixty-one years older than Rome. Was it? Since both cities were probably giving themselves the benefit of the doubt, who can tell?

The ancient Greeks, however, were flourishing when Rome was still an inconsequential town, and it can be supposed that Greek history can be pushed back reliably to a point considerably earlier than Roman history can.

The Greeks were not a unified people but consisted of scores and scores of independent city-states scattered all over the shores and islands of the Mediterranean and Black seas. Each city-state had its own customs, its own legends, its own ways—all contributing to the wonderful variety of Greek civilization, which some think was (despite its flaws) about the most attractive the world has seen.

There were three things the Greek cities held in common: the Greek language, the Homeric epics, and the Olympian games. The Olympian games took place every four years, and so important were they considered that even wars were suspended at the time of the games to allow them to be held in peace. (Nowadays, when there is a general war, it is the Olympic games that are suspended so that the war can be carried on undisturbed—just one way in which our civilization is less attractive than the ancient Greek.)

Eventually, Olympian games were used as a way of numbering the years. Years were counted by groups of four called *olympiads,* and each year was the first, second, third, or fourth of a particular olympiad.

When you have a particular event written up by different writers using different chronologies, you can match the chronologies. For instance, if Julius Caesar was assassinated in 709 A.U.C., according to a Roman writer, and in the first year of the 183rd olympiad, according to a Greek writer, then you can work out a formula for converting any Roman date to a Greek date and vice versa.

The Greek histories are judged to be quite accurate back to 600 B.C. (153 A.U.C.). Thus, Solon was made *archon* (ruler) of the city of Athens and set about reforming its legal system in 594 B.C.

It was about 750 B.C. that the Greeks picked up a system of writing from the Phoenicians; before that, there was only tradition. Later Greeks, working out history as far as they could, placed the year of the first Olympian games at 776 B.C. (twenty-three years before the founding of Rome).

The Trojan War, the subject of Homer's *Iliad,* may have taken place about 1200 B.C., but that doubtful date is as far back as we can push history by way of the Greeks.

But there were literate civilizations earlier than that of the Greeks. Since the Greeks obtained their writing from the Phoenicians and were in awe of the Egyptian and Babylonian cultures, those three must have been literate before the Greeks were.

The one source of ancient history known to the medieval students, besides Greek and Roman histories, was the Bible and that, too, seemed to indicate that Egyptian and Babylonian histories extended back far beyond Greek times.

There were written remnants of those histories, too. There were Egyptian inscriptions on the ancient structures and monuments that existed in that land. In addition, inscriptions incised in baked clay were found in Babylonia. The Egyptian writing was called *hieroglyphics* (from Greek words meaning "sacred carvings," because they were so often found in ancient temples). The Babylonian writing was *cuneiform* (from Latin words meaning "wedge-shaped,'' because the stylus forming the marks was so held as to make a wedgelike shape in the soft clay).

Undoubtedly, the hieroglyphic and cuneiform inscriptions could tell us a great deal about pre-Greek history, but the trouble was that whereas Latin and Greek were known to scholars, hieroglyphics and cuneiform were at first indecipherable and told the world nothing.

A turning point came when, in 1798, the French general Napoleon Bonaparte (1769–1821), in one of his more harebrained moments, led an expedition to Egypt in the face of superior British sea power. He managed to get his army to Egypt, and he eventually managed to return to France himself, but his army, for the most part, was left in Egypt, either dead or as British prisoners.

While his army was there, however, one of his engineers, whose name was Bouchard (or possibly Boussard—nothing else is known about him) came across a fragment of black basalt that was 45 inches long and 28.5 inches wide, with its corners knocked off. He found it near the Egyptian town of Rashid—''Rosetta'' to Europeans—thirty miles from Alexandria, so it came to be known as the Rosetta Stone.

On the stone was a thoroughly uninspiring inscription dated 196 B.C., that is, the ninth year of the Egyptian king, Ptolemy V (210–181 B.C.), thanking him for his help to the temples and the people. It was the typical slavering over a ruler to keep him in a good humor and to get still more money out of him.

The important thing, however, was that the inscription was repeated three times, once in Greek, once in Egyptian hieroglyphics, and once in Egyptian *demotic,* a simpler form of the hieroglyphics. It was assumed that each different form of writing carried the same message so that all the people of Egypt could understand it. Since the Greek message was perfectly plain to any scholar who knew Greek, the problem was to work out which Egyptian sign or signs corresponded to each of the Greek words. The Rosetta Stone, in short, was a kind of Greek-Egyptian dictionary, and deciphering of the hieroglyphics was at last possible. (Indeed, "Rosetta Stone" has entered the English language as a metaphor for any key to the understanding of some complex phenomenon that has hitherto been completely puzzling).

The decipherment of Egyptian was possible, but not easy. It took years to accomplish. The Rosetta Stone fell into British hands after the French in Egypt were forced to surrender; it was taken to the British Museum. There, scholars from all lands studied it and worked with it.

In 1802, a Swedish scholar, Johan David Akerblad, had the inspiration of turning to the Egyptians themselves. Egypt had been taken over by Islamic armies in 640, whereafter the Egyptians were slowly converted from Christianity to Islam and abandoned their own ancient language for Arabic.

But not entirely. There remained in Egypt a remnant of people who clung to Christianity. They are called Copts (a distortion of "Egypt"). The Coptic language is descended from the ancient Egyptian. Akerblad, making use of both the Greek inscription and the Coptic language, was able to translate a few phrases in the demotic portion of the Rosetta Stone.

In 1814, an English scholar, Thomas Young (1773–1829), took up the task. He decided that certain hieroglyphic signs in the Rosetta Stone, which were encircled in an oval as though they were especially important, must represent the names of the king and queen, Ptolemy and Cleopatra. Assuming this to be so (and it was so) he worked out the significance of several of the hieroglyphic symbols.

In 1821, the task was carried further forward by a French linguist, Jean Francois Champollion (1790–1832), who first realized that some of the hieroglyphic symbols represented letters, some syllables, and some words. It was an extraordinarily complicated language, but by the time Champollion was through, the worst of the task was accomplished. Later scholars worked out further details, and the whole world of Egyptian inscriptions opened itself to them.

A similar stroke of good fortune opened the cuneiform

writings to modern scholars. The Persian king Darius I (558–486 B.C.) had gained the throne in 521 B.C. by dubious means. As an exercise in public relations he put up an inscription on a cliffside near the now-ruined town of Behistun (or Bisitun) in what is today western Iran. It detailed the way in which Darius had succeeded to the throne (according to his own version of the affair). It was carved high on the cliffside so that it could be seen, yet not defaced. What's more, it was repeated in three cuneiform languages—Old Persian, Assyrian, and Elamite—so that as many people as possible of the polyglot empire could understand it.

Old Persian could be made out with the help of present-day Persian. Using that as a starting point, Assyrian and Elamite could be translated.

The decipherment was carried out by an English archaeologist, Henry Creswicke Rawlinson (1810–1895). To get at the inscription he had to dangle from a rope slung over the edge of the cliff, with the ground 500 feet below him. It took years for him to transcribe the message completely, but by 1847 he was busy deciphering the languages.

This eventually opened up all the cuneiform languages, and scholars could work out the long history of Mesopotamia —the valley of the Tigris and Euphrates rivers.

We know now that Egypt was most powerful under Thutmose III, who reigned from 1504 to 1450 B.C., almost three centuries before the Trojan War. The pyramids were built a thousand years earlier still, about 2400 B.C., and Egypt was first unified and made into a strong kingdom by Narmer about 2850 B.C. The stretch of time from the unification of Egypt to the Greek philosopher Socrates (470–399 B.C.) is equal to the time from Socrates to ourselves.

As for the Mesopotamian valley, before the Persian conquest it was ruled by the Chaldeans, of whom the most powerful monarch was Nebuchadnezzar, who ruled from 605 to 562 B.C. Before the Chaldeans were the Assyrians, who were most powerful under Esarhaddon, who ruled from 681 to 669 B.C. Long before that were the Babylonians, who flourished under Hammurabi, whose reign was from 1953 to 1913 B.C. The earliest of the great civilizations of the region were the Sumerians, who were at their peak under Sargon of Agade, reigning from 2360 to 2305 B.C.

It would seem, according to present thinking, that the art of writing was invented by the Sumerians about 3100 B.C., and that by 3000 B.C. the notion had spread to Elam in the east and to Egypt in the west. By 2200 B.C. it had spread to Crete, by 2000 B.C. to India, and by 1500 B.C. to the Hittites. China may have invented writing independently, but not until 1300 B.C. The Mayans in southern Mexico invented it, too, but not till 2,000 years later.

If, then, writing is the indispensable key to history, we can say that history began about 3100 B.C., that is, about 5,000 years ago. It began, however, in a small region near the mouths of the Tigris and Euphrates rivers in what is now southeastern Iraq. It spread out slowly, with new nuclei forming later in China and, later still, in southern Mexico. It is only in modern times that history has become worldwide.

Nevertheless, we must remember the principle of evolution. Before writing came into use there must have been some centuries of "pre-writing," a period when images or markings were made to guide human thinking. Thus, before the time of Columbus, the Incas of the Andean region of Peru did not have the art of writing, but they used an intricate system of colored cords with knots in them to record numerical information of various kinds. Writing was clearly on the way.

And even without writing, the Incas had a complex and smoothly working civilization. So must the Mayas have had before the development of writing. So must the Chinese, the Egyptians, and the Sumerians.

We might ask, then, when did civilization begin?

# 3

## CIVILIZATION

Until the last couple of centuries, the one source that the Christian world had for information on the earliest times of humanity was the Old Testament of the Bible. A good part of it was a set of documents dealing with the ritual and ethical details involved in the worship of the god, Yahveh. Since the chief worshipers were the people of Israel and Judah, there were historical sections dealing with those people and their immediate neighbors.

The historical sections are apparently taken from the secular annals of the time, and while they are overcast with the religious preoccupations of the writers, they seem to be reasonably accurate once the miracles and homilies are subtracted. Indeed the Book of Samuel and the Book of Kings may represent the earliest historical writings of quality that we have. They certainly antedate the works of the Greek "father of history" Herodotus (485–430 B.C.) by centuries.

The chief difficulty in dealing with Old Testament history is that it gives no dates in the modern sense—not one from beginning to end. It does give durations, however—

how long a particular king ruled, how old a person was when he gave birth to a son, how many years from this event to that. In addition, some biblical passages describe events that are dealt with by other historians who do give dates in chronologies whose relationship to our own we can work out.

This means that, starting with some firm dates, we can carefully work our way backward and perhaps come to the year of the events with which the Old Testament begins. One person who did this rather early on was an Irish-born Anglican bishop, James Ussher (1581–1656). As Varro searched and considered the early legends of Roman history and as Dionysius Exiguus considered biblical clues to the birth of Jesus, so Ussher traced his way back through the legendary tales in the book of Genesis. He calculated the probable times of all the events in the Bible, and these are included in many editions of the King James Bible (including the one I own).

About the earliest event in the Bible for which a date can be given with moderate assurance, from general historical considerations that are not dependent on the Bible alone, is the accession of Saul, the first king of Israel. The usual estimate is that this took place in or about 1020 B.C., when Egypt and Assyria were both going through periods of decline. That is why Saul's successor, David (1043–973 B.C.), was able to build up a realm including the entire eastern shore of the Mediterranean. As soon as Assyria recovered its strength, this brief moment of Israelite domination ended.

Ussher, however, gives the year of Saul's accession as 1095 B.C.

Before Saul, all is legendary, and there are no definite events attested to outside the Bible. There was, for instance, the period of the Judges, as given in the Book of Judges. The various Israelite tribes, in loose confederacy, had taken the land of Canaan (later called Palestine by the Greeks, from the Philistines who occupied the southeastern seacoast). The tribes fought against each other and, in consequence, were frequently under foreign domination, which might be ended by the emergence of a strong leader ("judge") in one tribe or another.

The Bible gives the lengths of time during which the various judges ruled, and by assuming that they did so one after the other, Ussher calculates that this period endured for 330 years and began in 1425 B.C. Modern biblical scholars feel that the judges ruled separate tribes and that their periods of domination probably overlapped. They estimate that the period of the judges may have lasted only 180 years and that it began about 1200 B.C.

Ussher places the conquest of Canaan itself, under the legendary Joshua, from 1451 to 1425 B.C. It is much more likely that it actually took place from 1230 to 1200 B.C., just before the Trojan War.

After all, between 1451 and 1425 B.C., the Egyptian Empire was still at its height and was in firm control of Canaan and surrounding regions. Desert tribes would not have had any chance of taking over any part of Canaan. Between 1230 and 1200 B.C., however, the Egyptian Empire had started a precipitous decline and the conquest would indeed have been possible.

Ussher places the Exodus from Egypt at 1491 B.C., but if it happened at all, it must have happened about 1237 B.C., at the end of the reign of Pharaoh Rameses II, when Egypt was having increasing trouble and was about to suffer the incursion of the "Peoples of the Sea" that reduced it nearly to chaos.

According to Ussher, the legendary Abraham arrived in Canaan in 2126 B.C. A few Christians, before the Roman Empire had adopted their religion, tried to establish a calendar that would show their history to be older than that of proud Rome and Greece, so they established the "Era of Abraham," dating the years from 2016 B.C., calculating his time at more than a century later than Ussher was to do. The Era of Abraham was never used by any considerable number of people.

The worldwide Flood was placed by Ussher in 2349 B.C., which was about the time Sargon of Agade was establishing his empire (without noticing any flood) and some time after the pyramids were built (with no record of any worldwide Flood appearing in the Egyptian records that continued, unruffled and unbroken, over that period).

In this case, Ussher was overconservative. There are signs of a huge flood in the Tigris-Euphrates valley (all river systems are given to flooding, as witness our own Missouri-Mississippi), but it took place about 2800 B.C. Of course, it was a local flood, confined to the valley, but it was so disastrous that the Sumerian survivors, appalled at the magnitude of the cataclysm in the only part of the world they knew, might well have considered it worldwide and reported it as such.

The flood struck a blow to the civilization of the period that was hard to surmount. It probably destroyed most of the records, and the Sumerians were left to invent extravagant legends for the pre-flood period—such as kings who reigned for tens of thousands of years and so on.

The earliest portions of the Bible were put into shape at the time when the Jews were captive in Babylonia (586–539 B.C.), and they picked up the Babylonian version of primeval history, including the story of the worldwide Flood.

Before the Flood, the Bible tells of the antediluvian patriarchs with extended life spans of nearly a thousand years each, a kind of timid echo of the listing of the Sumerian antediluvian kings with their far longer life spans. By tracing through the reported age of each patriarch at the birth of his oldest child, it is possible to come to the date at which Adam and Eve came into being and the creation took place.

Jewish scholars placed the date of creation in 3760 B.C., and the Jewish religious calendar counts the years from that time. This is called the Jewish Mundane Era, where *mundane* is from the Latin word for "world." A Mundane Era, in other words, counts the years from the creation of the world. This means that I am writing this sentence in the year 5847 of the Jewish Mundane Era.

Ussher calculates the date of the creation to be 4004 B.C., exactly 4,000 years before the birth of Jesus. (I doubt that this is a coincidence. I'm sure that Ussher adjusted

# RISE OF MODERN MAN

| | | |
|---|---|---|
| **HISTORIC TIMES** | 1987 | - - - - - - - - - - - - - - - - - - |
| | 1,000 | PRINTING WITH MOVABLE TYPE |
| | 0 (A.D. 1) | BIRTH OF CHRIST |
| | 1,000 B.C. | IRON AGE |
| | 2,000 B.C. | PYRAMIDS - - - - - - - - - - - - |
| | 3,000 B.C. | BRONZE AGE<br>Start Of History/Writing |
| | 4,004 B.C. | Year of Creation according to<br>Chronology of Bishop Ussher |
| **STONE AGE** / **Prehistory** | 10,000 B.C. | START OF CIVILIZATION<br>START OF AGRICULTURE |
| | 30,000 B.C. | CRO-MAGNON MAN - - - - - - - - |
| | 40,000 B.C. | HOMO SAPIENS SAPIENS —— MODERN MAN |
| | 100,000 B.C. | HOMO SAPIENS — NEANDERTHALENSIS - - + - - - |
| | | —————————————— HOMINIDS - |
| | 250,000 B.C. | |
| | 1.5 MYA* | HOMO ERECTUS —— JAVA MAN - - - - - - -<br>——— PEKING MAN |
| | | HOMO HABILIS    RISE OF MODERN MAN |
| **NEITHER APE NOR MAN** | 2 MYA | |
| | 3 MYA | |
| | 4 MYA | AUSTRALOPITHECUS ROBUSTUS · - - - - - - - -<br>└— AFRICANUS<br>└— AFARANSIS "Lucy" |

*Millions of Years Ago

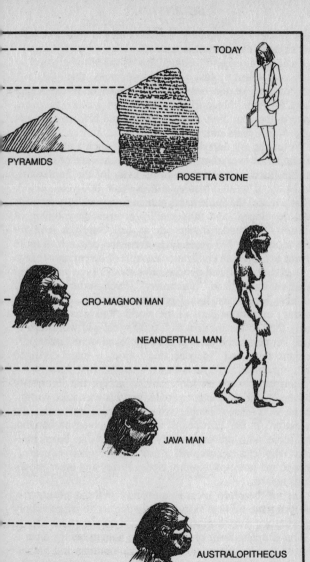

TODAY

PYRAMIDS

ROSETTA STONE

CRO-MAGNON MAN

NEANDERTHAL MAN

JAVA MAN

AUSTRALOPITHECUS

some of the less calculable dates in order to come out with just such an even result.)

Until the nineteenth century, it was taken for granted by Christians, even by historians and scientists, that 4004 B.C. was the date of the origin of the Universe. If we accept this Christian Mundane Era, then I am writing this sentence in the year 5990, and the world and Universe are not quite 6,000 years old.

We might ask ourselves now whether such a date for the beginning of everything is plausible on the face of it. After all, the historical record we have, even for the Sumerians, places all of written history within this 6,000-year period. What's more, the Bible treats human beings as fully formed, fully developed, and fully intelligent from the moment of creation—and under divine care as well. Surely it need not take more than 900 years to go from the origin to a quite advanced Sumerian civilization capable of inventing writing.

Of course, civilized peoples have always been surrounded by as yet uncivilized "barbarians." Even in the nineteenth century, Europeans found primitive peoples, who lacked writing, in various parts of the world. This was not necessarily fatal to the thought of a 6,000-year-old world, however. Perhaps some parts of the population were "inferior"; perhaps they had "degenerated" from a more civilized past.

Europeans were all too ready to accept the inferiority and degeneration of other people, but this was quite wrong. There were many people who were civilized when the ancestors of the Europeans were themselves barbarians, and those who are supposedly barbarians can beget children who can accomplish great things when educated, even to the point of winning Nobel Prizes and other prestigious awards.

Let us therefore look at humanity without necessarily accepting the biblical story literally and try to judge simply by what we can observe and deduce.

The simplest form of organization human beings have is that of family groups who subsist by hunting and gathering, by tracking down and killing small edible animals, and by gathering edible plants. This is the precarious kind of life that all animals live.

Human beings, even in prehistoric times, must have been much more intelligent than other animals, and this surely must have helped them in their hunting and gathering, but it was a way of life that was still precarious even so. It is sometimes estimated that the Earth could not support more than a total of 20 million people who live by hunting and gathering alone.

Even today there are some primitives who live in this way, but most human beings now live in more complicated fashion. At some time in the past, groups of people must have learned to roast grain to make the ears edible, and then to grow such grain deliberately so as to have a large supply of food always handy. People learned to tame animals, keep them under control, and encourage their breeding, so as to have a regular supply of meat, milk, eggs, hides, wool, and other useful commodities.

In short, people developed agriculture and herding. This made it possible to extract a much larger supply of food from a given acreage of land, and the population naturally increased.

For the first time in history, in fact, there was the possibility of having more food than was needed so that some people didn't have to be engaged in growing food but could do other things—make tools, for instance, or tell stories, in exchange for food.

The population, in short, not only increased but became specialized.

There was, however, a price to be paid. Hunters and gatherers are free to move about; indeed, they *must* move about, for if they stay in one place too long, they will consume all the food the region has to offer. Those who herd, however, are tied to their flocks and cannot stray from them. Those who farm can't move at all for they must remain in the vicinity of their immovable crops.

Furthermore, they must protect their supply of food from hunters and gatherers who would be delighted to help themselves to the unusual supply collected, through ardu-

ous labor, by herders and farmers. The herders and farmers have to gather in fixed places, then, in close proximity
to each other so that they can cooperate in defense. They
have to choose a good site with an assured water supply,
situated on a height if possible, or behind walls, to make
the defense easier.

What's more, the new way of life calls for foresight, for
willingness to work very hard for months, without immediate reward, but with expectation of a useful harvest
eventually. It also requires cooperation between individuals and groups since, generally, the harvest cannot be
assured without irrigation from some nearby river, and
since irrigation won't work unless a system of trenches and
dikes are built and kept in constant repair.

To ensure this cooperation and arrange for the making of
decisions, groups of people must choose rulers, both secular and spiritual (sometimes both together), or have rulers
forced upon them. They must maintain soldiers and they
must pay taxes. In short, an agricultural and herding society
is *much* more complex than a hunting and gathering society.

An agricultural and herding society is, on the whole,
more secure and offers more variety, but there are always
those who hark back to the idealized simplicity of hunting
and gathering—hence the legendary "golden ages" with
which people populate the past and, in particular, the tale
of Adam and Eve gathering fruit idyllically in Eden until
they are ejected to face a life of farming and herding
because they had learned too much.

In any case, the mark of the new society was the city,
very small and simple at first, but growing larger and more
complicated as population grew and wealth accumulated.
The Latin word *civis* means "city" and a *civitas* is a "city
dweller" or "citizen." When people gather in cities, then,
they are *civilized* and represent a *civilization*.

Civilization does not necessarily involve writing, but it
makes writing unavoidable, eventually. As civilization grows
more complicated a system of writing becomes necessary,
if only to keep records of crop production, to calculate

taxes, to list receipts, to send messages back and forth assuring cooperation, and so on.

Every society that has developed writing has been a civilization at the time, even a fairly advanced one. Hunting and gathering societies are too simple to require writing, and societies do not go to the great trouble of developing a writing system until driven to it.

In that case, we must assume that the Sumerians, in inventing writing in 3100 B.C., must first have adopted agriculture and herding, developed a system of irrigation in the Tigris-Euphrates valley, developed governments that handled both secular and religious matters (successful agriculture required, in the view of early farmers, a great deal of propitiation of the capricious gods), a trained army with armor, weapons of war, carts for transportation, and so on.

All this takes time. No Sumerian woke up one day and said, "Oh, my, it has just occurred to me to grow grain for a harvest. Let's start working out an irrigation system."

Instead, it would all develop by innumerable small steps, by fits and starts, by attempts and failures. What it amounts to is that the 900 years that stretches between Ussher's date of creation and the Sumerian invention of writing *is not enough*. We cannot expect that in 900 years a civilization complex enough to force a writing system on people can be developed.

This is obvious to us since we know how slowly and uncertainly any evolutionary policy proceeds. (Think how long it took the people of the United States to do something as decent as to abolish slavery, or, since then, how much time it is taking to do something as decent as to avoid judging people by their complexions or accents.)

Evolutionary slowness was also, I believe, obvious to the ancients. All the ancients seemed to think that not only were human beings created by gods, but that they were also given civilization by gods. There simply didn't seem time, or enough human ability, or both, to do it without divine help.

Thus, in the Greek myths, Prometheus stole fire from the Sun and gave it to human beings; Athena gave humanity the secret of olive cultivation and the art of weaving; Demeter, the technique of farming; Poseidon, the war-horse; Apollo, the arts; and so on.

In the Bible, Adam's oldest son, Cain, was already a farmer, and his second son, Abel, a herdsman. How did they learn agriculture and herding? The Bible doesn't say so, but it seems plain that God must have taught them this.

Even in today's secular age, it seems difficult to believe that ancient peoples accomplished what they did all by themselves. How did the Egyptians build the mighty pyramids with scarcely any technology worth mentioning? If we are too sophisticated to accept gods or demons, then we may seek a "scientific" equivalent—intelligent beings from outer space.

In recent years, books about such "ancient astronauts" have made their authors rich, despite the fact that they are totally empty of significant content.

Such theories, whether involving gods, demons, or ancient astronauts, are insulting to the indomitable spirit of humanity. *People* accomplished the establishment of civilization and all it led to, and they should not be deprived of the credit. The Egyptians *did* build the pyramids, and they did it by spending many centuries in developing techniques for the purpose and in building first very simple forerunners of pyramids, then more complicated ones, and so on. Finally, they learned how to build the full-sized pyramids.

We must conclude that the prehistoric period of human development must extend back beyond 4004 B.C., perhaps even far beyond it. But, without writing, how can we find out how far back in time such prehistory extends? As I said earlier, without writing we can't learn much about specific events, but we can learn some general facts.

The study of prehistoric times is called *archeology*, from Greek words meaning "the study of ancient things."

People have always been interested in human-made objects from the past. In Great Britain, for instance, people were interested in finding and studying Roman relics—old spear-points, or coins, or pieces of pottery. Such people were called *antiquarians,* and it was a respectable and harmless study.

It became something more serious in the eighteenth century, in connection with the old Roman cities of Pompeii and Herculaneum. These towns, just south of Mount Vesuvius, flourished in the first century of the Roman Empire and had no sense of doom about them because Mount Vesuvius had not been active in human memory. However, volcanoes believed dead can come back to life, and on August 24, 79, Vesuvius went into action with a roar and buried the cities. Pompeii was covered with 20 feet of ash and debris; Herculaneum was buried deeper still.

In 1709, and periodically thereafter, people began to dig into the mound that covered Pompeii, and all sorts of artifacts were uncovered: statues, pottery, remnants of houses, furniture, inscriptions. In short, Pompeii turned out to be a rich storehouse of information about everyday Roman life that could not be found in Roman histories.

That was the first realization that Europe had as to the usefulness of digging into ancient ruins. If anything more was needed, it was the career of a German merchant named Heinrich Schliemann (1822–1890). From childhood, he had been fascinated by the story of Troy as told in the *Iliad,* and he was firmly convinced that the story was not a myth, but (once the gods were omitted) true. It was his obsession to find traces of the city. He worked with incredible dedication to make himself rich and succeeded in doing so.

In 1868, he finally went east and began his research. Going by the descriptions in the *Iliad,* he decided that a mound at the little town of Hissarlik in northwestern Turkey must be the site of Troy, and in this he was apparently correct. He dug into the mound enthusiastically, but unscientifically, in order to get to the lowest levels (unnecessarily destroying much in the upper levels). He located a city he identified with Troy, and he also located cities earlier still.

He made important findings in the ruins of Mycenae in

mainland Greece. This was the most important town in Greece at the time of the Trojan War and was the home of Agamemnon, the Greek commander.

Schliemann showed that there had indeed been a Bronze Age civilization in Greece (one in which the smelting of iron ore had not yet been developed) and that Homer had described it with surprising accuracy. This Homeric civilization antedated the known period of classical Greece, and eventually this led to the discovery of the Minoan civilization of Crete, which was flourishing as early as 3000 B.C. with elaborate buildings and (this never ceases to impress me) internal plumbing.

Crete was the first civilization to develop a navy (it was an island, after all), and so efficient was the navy in protecting its shores that its cities lived unwalled and in peace. When the Minoan civilization was destroyed, about 1400 B.C., it was largely the result of a volcanic eruption on an Aegean island to the north. Crete suffered catastrophic destruction through the fall of ash and the slamming of tsunamis (so-called tidal waves) into its shores.

Schliemann's findings made an enormous impression on the world, not only because of his discoveries in themselves but because they involved the Trojan War, which, for 2,500 years, had permeated the consciousness of the Western world, thanks to the supreme artistry of Homer.

Everywhere, ancient ruins began to be probed by methods that grew to be much more careful, painstaking, and scientific than anything Schliemann had done.

The Hittite civilization was revealed in Asia Minor. From references in the Bible, the Hittites were thought to be a very minor people in Canaan, but it turned out they were, in their time, a powerful empire, which, about 1350 B.C., fought by the Egyptian Empire at its peak to a standstill.

Toward the end of the nineteenth century, details of the Sumerian civilization, the oldest on Earth, were first uncovered, and between 1922 and 1934 the English archeologist Charles Leonard Woolley (1880–1960) worked out virtually its entire history by excavations at the site of the ancient city of Ur (from which Abraham, according to the biblical account, emigrated to Canaan).

But if one picked up an artifact from a ruin, how old might that artifact be if no date of any kind was found on it?

The simplest way of dating an artifact is by considering its position. An artifact is usually found buried at some depth beneath the surface. In general, one can suppose that objects that are found at the same depth are of the same age, while objects deeper than other objects are older. This is by no means a sure thing, for it sometimes happens that the sites may be mixed up by either natural processes or human ones.

There are various other ways of judging relative age, and in the end, after considerable detective work and careful reasoning, the artifacts in a particular dig can be lined up pretty reliably in order of increasing age.

What's more, you sometimes find an object manufactured in a distant region present among locally manufactured objects (after all, trade existed even in very ancient times). You can then *crossdate*. If you know the relative date of the foreign object, you can assume that the local objects are about as old. This is particularly useful if the foreign object is from a civilization with writing, while the local artifacts are not. You might then have an absolute date for the foreign object and apply it to the local ones. However, absolute dates by cross-dating can't go back beyond 3100 B.C. Is it possible to get absolute dates earlier than that in some other way?

Surprisingly, the answer is yes.

For instance, in some cases, sediment is deposited in lakes in a periodic manner. Each winter, a fine, dark sediment is laid down, but in the spring and summer, when snow and ice are melting, a coarser, light sediment is brought down. One can study the sediment and count the layers, knowing that each light-dark layer stands for one year. Such repetitive layers are called *varves,* from a Swedish word meaning "periodic repetition," for it was in Sweden that these phenomena were first noted in glacial lakes.

By extension, the term *varve* can be applied to any other regular layering of sediments—as a result of periodic drying or periodic wind changes, and so on. The first person to try to establish actual dates by this means, and to date artifacts found in such sediments, was a Swedish geologist, Gerard de Geer (1858–1943). By now it is possible to count back through 18,000 years by means of varves, and that is sufficient, by itself, to make mincemeat of Bishop Ussher's notion of a 6,000-year-old world.

Then, too, an American astronomer, Andrew Ellicott Douglass (1867–1962), who worked in Arizona, began to study wood. Old pieces of wood were perfectly preserved in Arizona's dry climate, and what he studied were the tree rings.

Every summer, wood will generally grow rapidly if the weather is suitable over the year; slowly, if it is not. This pattern of rapid and slow growth produces the effect of rings, one ring for each year. If a summer is unusually cool or unusually dry, the growth ring is narrow. A warm, wet summer, on the other hand, produces a wide growth ring.

In a living tree, Douglass would find a particular pattern of rings, wide and narrow, that might extend back a hundred years. (It is not necessary to kill the tree to do this. A core of wood can be bored from back to center, taken out, and studied. The tree will heal).

Suppose you studied a piece of wood that you suspected was part of a tree cut down a few decades before. Its ring pattern would fit an older portion of the pattern of the living tree, and counting back to the place where the pattern began to fit, you might find the wood came from a tree that was cut down as many as thirty-four years ago, and you could follow the pattern farther back than the original pattern you were dealing with.

A still older piece could be matched against the older pattern, and the pattern pushed still further back. By 1920, Douglass had worked out a pattern that stretched back to about 1300 A.D. That meant one didn't need human dating. If an ancient Indian village was discovered, the wood used in constructing a house would give the date of the house from the tree-ring pattern. Later work carried the tree-ring pattern back for 8,000 years.

Such dating methods are rather specialized and can't always be applied. Something much better was recently developed.

In 1940, a Canadian-American biochemist, Martin David Kamen (b. 1913) isolated a variety of carbon called *carbon-14*. Carbon-14 is radioactive and breaks down slowly and very regularly at a rate such that half of any quantity is gone in 5,700 years. Half of what is left is gone in another 5,700 years, and so on. In breaking down, carbon-14 gives off subatomic particles that can be detected with great delicacy so that the breakdown rate can be precisely followed.

Even at this slow rate of breakdown (slow in terms of human lifetimes), any quantity of carbon-14 that might have existed on Earth when it came into existence would be long gone. (We'll talk about the age of the Earth in a later chapter). Nevertheless, carbon-14 exists in the atmosphere right now, because it is constantly being made. Cosmic rays from outer space smash into atoms in the atmosphere and produce a certain small amount of carbon-14. The production just balances the breakdown so that the level remains constant.

Plants absorb carbon dioxide from the air, and some of the carbon dioxide contains carbon-14 atoms, which become part of the plant tissue. Those carbon-14 atoms are regularly breaking down, but new carbon-14 keeps being introduced. The absorption and breakdown balance to leave a certain fixed level of carbon-14 in all living plants.

Once a plant dies, the carbon-14 in its tissues continues to break down, but no new carbon-14 is added. For that reason, one can tell how long a plant product has been dead by the amount of carbon-14 left in it, and this level can be determined by measuring the amount of subatomic particles of a particular type being given off.

In this way, wood, textiles, pieces of charcoal from campfires, or anything organic can be dated. The American chemist Willard Frank Libby (1908–1980) perfected the technique in 1947, and since then objects up to 45,000 years old have been dated.

By the use of carbon-14 dating techniques, for instance, there seem to be traces of farming and of habitations on the site of the town of Jericho as long ago as 9000 B.C. —almost 6,000 years before writing was invented anywhere. There may be places where agriculture started a thousand years earlier still, so we can say that civilization is 12,000 years old or just twice as old as Ussher thought the Earth and the universe were.

Of course, even before civilization began, human beings existed as hunters and gatherers and were just as intelligent, individually, as civilized human beings are today. We might ask, then, whether human beings had a beginning, too, one that would be, of course, older than the beginning of civilization. To narrow down the question we might ask about the beginnings of "human beings like us," and refer to such beings as *modern man*.

This is not an entirely happy phrase these days. *Man* has long been taken to have two meanings. One is general, and refers to all human beings, male and female, adults and children. The other is particular and is applied to adult males only. This is unfortunate, for when *man* is used in the general sense, it can be taken, by sensitive women (or children), to exclude them from consideration, as though they were not quite human.

One can understand the sensitivity, and these days I try to use such words and phrases as *person, people, human beings,* and so on in place of *man*, unless I *do* mean an adult male. In this case, however, *modern man* is so commonly used to mean "human beings like us" that I feel I have no alternative but to use it.

# 4

## MODERN MAN

The tools that archeologists discover are made of different materials. In any given region, tools that are of a comparatively recent manufacture may be made of iron. Tools that are older are often made of bronze. Tools that are older still are made of stone.

This is no mystery. Stone has always been around, but bronze has to be smelted from certain mixtures of copper and tin ores, a comparatively advanced technology that took a long time to work out. Iron must be smelted out of iron ore, which is more common than copper and tin ores, but such smelting requires more heat and is a trickier technique.

In 1834 the Danish archeologist Christian Jurgensen Thomsen (1788–1865) first divided human history into a Stone Age, a Bronze Age, and an Iron Age.

In different regions, these ages are found at different times. There are a few isolated places where people are still in the Stone Age, but most civilizations are by now in the Iron Age, because they either worked out iron smelting for themselves, borrowed it from neighbors, or had it brought to them by conquerors.

In western Asia, where civilization is oldest, the Bronze Age may have begun by 3000 B.C. and the Iron Age by 1300 B.C. Both the Bronze Age and Iron Age, then, are essentially periods in historic times. Before 3000 B.C., that is, in prehistoric times, the whole world was in the Stone Age.

It was eventually recognized, however, that the Stone Age was by no means a uniform period. There was a slow increase in sophistication in the manner in which the stone tools were produced, and the rate of increase itself increased with time. (This is a characteristic of technology that has been continuing to the present time).

In the last few thousand years before the coming of the Bronze Age, stone tools were formed by grinding and polishing, rather than by chipping. A British archeologist, John Lubbock (1834–1913), suggested in 1865 that the last few thousand years of the Stone Age be called the New Stone Age, or in Latin, the *Neolithic*. That would be the age of polished stone tools. Everything before that would be the Old Stone Age, or, in Latin, the *Paleolithic*. That would be the age of the chipped stone tools.

It was at the beginning of the Neolithic period that agriculture and herding came into use, that cities began to exist, that civilization started, and that the first "population explosion" resulted. This is sometimes referred to as the *Neolithic Revolution*. If, then, we talk about human beings as they existed before the Neolithic Revolution, and before civilization began, we are talking about *Paleolithic man*. How far back can we trace him?

To begin with, it is necessary to explain that all human beings on Earth, however different they may appear superficially, are essentially alike. Humanity today forms a single species and can interbreed freely. Differences in color of hair, skin, and eyes are largely due to differences in the quantity of a pigment called *melanin*, and this does not affect humanity's essentially unitary character. Nor do differences in the shape of the eye or nose, in the shape of the skull, or in height.

To be sure, these have all made enormous differences in

history and in social and psychological reactions, but that does not make them biologically important. The tragedies that are an outgrowth of the noted differences in human varieties are more an expression of psychopathology than of biology. After all, the same tragedies can arise out of a difference in religion, and there is no one who will claim that *that* represents a biological difference.

The Australian aborigines and the American Indians, who alone occupied Australia and the American continents, respectively, before the Europeans came, are as much modern man as are the proudest Europeans.

Both in Australia and the Americas, one can unearth burial places and find skeletons of human beings who died before, even long before, the arrival of Europeans. All the human bones ever found in either Australia or the Americas are those of modern man. They do not differ significantly from each other or from us. There are individual variations, as there are among living human beings (differences clear enough to let us distinguish one friend's face from that of another at once, without giving rise to any suggestion that any one of them is anything but completely human). There are also variations due to sex and age, or those imposed by diseases that affect the bones, such as arthritis or rickets. There is nothing systematic, however, that will mark any of the skeletons as a species that is not modern man.

What's more, if early American and Australian skeletons are dated by any of the methods available to archeologists, it is clear that none are older than a certain maximum age. The conclusion is that at some time in the past Australia and the Americas were totally uninhabited by human beings—until, at some point, modern man arrived from elsewhere and colonized those empty continents. (The same is true of almost all the islands of the world.)

Most archeologists are convinced that human beings entered North America from northeastern Siberia. This, naturally, would have to be at a time when sea level was considerably lower than it is now because so much water was tied up in the huge ice caps that rested on northern

Siberia and North America during the Ice Age. The lowering of sea level meant that there was a broad bridge of dry land between Siberia and Alaska, at least until the glaciers melted.

Skirting to the south of the glaciers, human beings crossed this land bridge, settled in North America, and gradually worked their way down into Central and South America.

At much the same time, human beings in southeast Asia took advantage of the lowered sea level to cross from the western Indonesian islands into New Guinea, then into Australia, and finally Tasmania.

In both cases, the migrations seem to have started about 25,000 to 30,000 years ago. It was not until about 8000 B.C. that human beings reached the southern tip of South America and perhaps not until A.D. 1000 that human beings first reached New Zealand.

We can conclude, then, that modern man must be at least 30,000 years old, for the first human beings who entered Australia and the Americas were undoubtedly modern man.

Prior to 30,000 years ago, all human beings alive on Earth must have lived in Europe, Asia, Africa, or on some of the islands near the continental shores. The question, then, is, when did modern man come into being in this large land mass that is sometimes referred to as the Old World and sometimes as the World Island.

In 1868, a number of human skeletons were found in a cave named Cro-Magnon, which is about seventy-five miles east of Bordeaux in France. They are representative of what is now called *Cro-Magnon man*. Other such remains over 30,000 years old have also been discovered.

Tracing modern man farther back is very difficult, and his appearance seems to be a relatively sudden one. We can't be sure when and where modern man first appeared, but the usual estimate places man at about 40,000 years ago.

We'll have to look further, but let's not insist on "modern man." The scientific name assigned to modern man is *Homo sapiens* (Latin for "man, the wise," which may be a bit of unjustified self-praise). Can there be older varieties of Homo sapiens that are not quite modern man?

# 5

## HOMO SAPIENS

If we were to decide that modern man began, quite suddenly, some 40,000 years ago, it need not necessarily bother those who would prefer to accept the biblical account.

Accoding to Genesis 1:26–27, "And God said, Let us make man in our image, after our likeness . . . So God created man in his own image, in the image of God created he him. . . ."

In Genesis 2:7, in a second account of creation, the Bible says, "And the Lord God formed man of the dust of the ground and breathed into his nostrils the breath of life; and man became a living soul."

Either way, whether God merely expressed his will, or whether he actually formed a human being of clay as a potter forms a vessel, one moment human beings did not exist, and the next moment they did.

Although Bishop Ussher calculated that this creation took place in 4004 B.C., his calculations are not the word of the Bible. The Bible itself does not give the time; it does not say how long each day of creation was, it does not say how long the primeval years were or whether there were any gaps in the record. If modern man came suddenly into existence 40,000 years ago, as archeological evidence seems to show, then might that not still fit the biblical account?

Yet there is an alternative: evolution. Human technology and the human social system evolved, rather than sprang fully developed into existence. Might that not be true of humanity itself as well? Might it be that modern man did not appear suddenly, but rather as the result of an accumulation of small changes—developing, in this way, from living things that were not themselves quite modern man.

That might seem to be stretching analogy too far. Until now we have been talking about mechanical and social phenomena. Whatever it is that has evolved, whether airplanes or civilization, has done so under the guiding direction of the human mind. If, then, human beings have themselves evolved and developed out of something less complex and advanced than a human being, what was the guiding mind that produced that change?

We might answer, "God!" but the Bible doesn't allow that as an answer. It says, instead, in Genesis 1:11, "And God said, Let the earth bring forth grass, the herb yielding seed, and the fruit tree yielding fruit after his kind . . . and it was so." Then, in Genesis 1:21, "And God created great whales, and every living creature that moveth, which the water brought forth abundantly, after their kind, and every winged fowl after his kind. . . ." Then, in Genesis 1:24, "And God said, Let the earth bring forth the living creature after his kind, cattle, and creeping thing, and beast of the earth after his kind: and it was so."

There could be more argument as to whether the biblical word *kind* means what the scientist means when he says "species," but there can't be any argument that the Bible says that the various kinds of plant and animal life were created as *different* kinds. From the very moment they were created, they existed separately, and there would seem to be no question of one changing into the other—a dog into a cat or a giraffe into an oak tree.

What's more, our own observations seem to jibe with

this interpretation of the biblical statements. Cats give birth to cats, while dogs give birth to dogs. There are no cases of dogs producing cats or of cats producing dogs. Furthermore, if we consider ancient descriptions of certain animals, or see ancient art featuring those animals, there is no question but that our animals are their animals, and without change.

Yet the apparently unlikely suggestion of biological evolution would not be squashed.

For one thing, life can be classified in a neat way. There are doglike animals (foxes, wolves, jackals, coyotes) and catlike animals (tigers, lions, leopards, jaguars). There are cattlelike animals (bisons, buffalos, yaks). There are horselike animals (donkeys, mules, zebras). The doglikes and the catlikes are alike in being carnivorous. The cattlelikes and the horselikes are alike in being herbivorous. All the ones I have mentioned are alike in having hair and in bearing live young that feed on milk.

There are birds, reptiles, and fish, each quite different from the others, but alike in having internal skeletons of similar composition.

In fact, it is possible to arrange life into a treelike affair, a trunk labeled "Life" that branches into plants and animals, each of which branches into large groups, which in turn branch into smaller groups, then into still-smaller groups, until finally one has tiny twigs branching into the twiglets that represent all the different species of living things. (There are at least two million different species now known, most of them insects, and there may be millions more that remain to be discovered, again most of them insects.)

Many people have tried to arrange such trees of life. Even I, at the age of ten, when I was fiercely reading natural history, tried to draw one, convinced that I had thought of something original, but quickly abandoned it when it grew too complicated for me to manage.

The first to make a really successful classification of living things was a Swedish botanist, Carolus Linnaeus (1707–1778). In 1735, Linnaeus classified plants in a par-

ticularly methodical way. He began by classifying similar
*species* into *genera* (singular *genus*), similar genera into
*orders,* similar orders into *classes* and so on. In 1758, he
extended the system to animals. What's more, he originated the notion of referring to each different form of life
by the name of the genus and the species, the two final
divisions. It was he who first classified humanity as Homo
sapiens, for instance.

The fact that the classification of living things somewhat
resembled a tree could not help but suggest to some people
that the tree of life grew like a real tree. Perhaps, originally, there was one simple form of life that with time,
split into two types, which further split and further split
until finally they had split into the twiglets that represented
the single species, doing it all in tiny steps that took an
enormous amount of time.

This seemed to make sense: If the various forms of life
had been created independently (either as described in the
Bible or in any other way) it would seem that there would
be no necessary connection among them. Why should they
exist in groups, and in groups of groups, and in groups of
groups of groups, and so on? Independent creation wouldn't
do that, but biological evolution would.

Such an argument is suggestive, but not compelling.
Linnaeus and some who followed him in extending and
further refining the scheme of classification did not accept
biological evolution.

One can easily present three arguments against biological evolution. First, if it accounted for the diversity of life,
then it should still be going on, and anyone can see that it
*isn't* going on. Second, God is perfectly capable of creating life in a related system of groups and groups of groups
for his own purposes. Third, even if evolution were considered to be taking place, there would have to be a
guiding intelligence behind such evolution and that would
have to be God, but the Bible denies God's use of evolution in creating life.

The answer of the evolutionists to the first argument is
that biological evolution goes on so slowly that it isn't

visible to the naked eye, so to speak. Nothing may be visible in the thousands of years of civilization, but we may be talking about millions of years.

This is not a sensible argument to people who are convinced that, in line with the Bible, the Earth is only 6,000 years old. Still, as the nineteenth century progressed, the arguments in favor of a great age to the Earth grew more powerful and convincing, as we shall see in later chapters.

The second point about God doing whatever he wishes is unanswerable, but it is the kind of argument that is not permitted in science. Anyone faced with any problem can shrug and say, "It is God's will," and if that is admitted as a permissible statement then all science comes to an end.

The third point about the need for a guiding intelligence is hard to answer. Those who thought that biological evolution took place had difficulty in pinpointing a mechanism that would make it work without calling upon a guiding, divine intelligence.

The best-known form of the argument is this: If you were to find a watch in the desert, perfectly made and running accurately, you wouldn't assume that it had just formed itself spontaneously. You would assume it had been formed by some intelligent being, presumably a human being, who had left it there for some reason. There would be no question about that.

Well, then, if you see the Universe and everything in it, infinitely more complex than a watch, and working with infinitely more precision, must you not likewise assume an intelligent being as its creator, a being as much more intelligent than a man as the Universe is more wonderful than a watch—in short, God?

To those who would not accept evolution, this seemed to be an absolutely unanswerable argument, yet it was answered. The English naturalist Charles Robert Darwin (1809–1882), after years of study and thought, published a book in 1859, the title of which is *On the Origin of Species by Means of Natural Selection*.

That last phrase is the key. As species reproduced themselves, there would always be small variations among the new generation, variations in size, in strength, in shape, in behavior, in intelligence, in endurance—in any of innumerable qualities. So far all would be random. However, some variations would better suit the species to the environment, and on the whole those variations would better survive. They would be "selected" by the influence of their natural environment. Natural selection would not act through intelligence, but the results that followed would be the same as though it *did* act through intelligence.

In the century and a quarter since that book was published, enormous advances have been made in many fields, advances that have served to refine and strengthen Darwin's thesis. The result is that biologists today accept biological evolution as a fact—even as the central fact of biology—although there is still vigorous argument over details of its mechanism.

Therefore, in searching for the origin of modern man, we must ask ourselves not only when and where modern man appeared, but from what creature, not quite modern man, did modern man *evolve*. For that, let us backtrack a little.

Dawin's explanation of the driving force behind biological evolution did not rest on philosophical argument alone. That would only make it reasonable. To make it compelling (to *force* acceptance even against one's will) there must be evidence. Such evidence existed before Darwin wrote the book, and much additional evidence supporting evolution, in many fields, has been discovered in the time since Darwin. (To be sure, there are people called "creationists," who insist, even today, on the literal words of Genesis and who argue against evolution. Their arguments are totally devoid of intellectual content, however, so we need waste no time on them.)

One of the most important strands of evidence supporting evolution (and certainly that which is best known to the general public) consists of the fossils that have been discovered. *Fossil* is from a Latin word meaning "something that is dug out of the ground.'' That word came to be applied particularly to those things, dug from the ground, that had a resemblance to living organisms or to parts of living organisms.

Such fossils were noted even in ancient times, but most people didn't know what to make of them. There were suggestions that they were just freaks of nature or that they were part of a life force that made even rocks strive to bring forth something with the appearance of life. During the Middle Ages, there were suggestions that fossils were Satan's attempt to imitate the work of God in creating life, and of course Satan failed miserably. Others held that perhaps God had tried making life until he was sure he had it right and that the fossils were his practice shots, so to speak.

Leonardo da Vinci was the first to advance a reasonable explanation. He thought that fossils were the remains of objects that were once living organisms. These had somehow gotten buried in mud, and slowly the composition of their bodies was replaced by a rocky substance until they were finally stony duplicates of the flesh and blood original.

The English naturalist John Ray (1627–1705) took another step forward. He was attempting a classification of plants and animals (and his work was the best there was before the time of Linnaeus), so he looked at the fossils from that standpoint. He noted that while fossils resembled living organisms, the resemblance was not complete. It was as though they represented organisms that were related to certain living organisms but were not identical to them.

He suggested in 1691 that fossils were, by and large, the remnants of ancient plants and animals that were not like those living in the present day and that they no longer existed today because they had become extinct.

The notion that a living thing could become extinct argued against the perfection of God's creation so that Ray's view was not accepted (and he was quite nervous about advancing it in the first place). Still, as more and more different fossils were found, Ray's view came to seem increasingly likely.

In order to avoid having fossils make it appear that the Earth had lasted a long time and that some species became extinct while others flourished (all of which would seem to encourage ideas of evolution), a Swiss naturalist, Charles Bonnet (1720–1792), suggested that fossils might represent life forms that had died in Noah's Flood and that had become extinct in that way.

In 1770, in fact, he generalized this notion and suggested that there were a whole series of catastrophes in which life on Earth was entirely wiped out and a new creation begun. The Bible, he argued, dealt only with the Earth after the last catastrophe, and it described a still later catastrophe (Noah's Flood) that was not quite total.

This point of view, called *catastrophism,* has had a kind of rebirth lately, but in the form presented by Bonnet it would not stand up. As the fossil record grew, more and more catastrophes had to be called on, and it was increasingly clear that no catastrophe had succeeded in wiping out all life. Fossils came more and more to imply evolution rather than catastrophe. (Bonnet was the first to use the word *evolution* in this connection, by the way.)

The matter of fossils sprang into particular prominence through the work of an English geologist, William Smith (1769–1839). It was a time when the English countryside was being cut into, here and there, in order to create canals for transportation. Smith surveyed canal routes and traveled over the country to study canals. He became interested in the layers of rocks exposed by the cuttings. These layers were sometimes sharply distinct from each other. Such layers were called, in Latin, *strata* (singular, *stratum*), and that is what they are now called in English, too.

By 1799 he had begun writing on the subject, and his enthusiasm was so long-continued and all-embracing that he became widely known as Strata Smith. His key observation was that each stratum had its own characteristic type of fossils not found in other strata. No matter how the strata were bent and crumpled—even when one sank out of view and cropped up again miles away—its fossil content remained characteristic of itself. In fact, it was possible to identify a particular stratum that one had not observed previously simply through its fossil content, a point Smith made in 1816.

It was possible to arrange strata in a regular series from those nearest the surface to those deepest. If we assume that each stratum consists of mud, or sediment, deposited out of water and that this sediment has been converted by heat and pressure to *sedimentary rock,* it makes sense to suppose that the deeper a stratum, the older it is.

It further appeared that the deeper a stratum, the less the fossils in it resembled living forms of life. If one works from the oldest strata toward the youngest, one can see life forms change slowly but surely in the direction of modern life. It is almost like watching evolution take place before one's eyes.

Naturally, the record is not complete. Even today, known fossils represent only about 200,000 different species of life, and this cannot be more than 1 percent of the total. In Smith's day the number of different fossils known was far fewer.

The reason for this paucity of fossil remnants is that in order for a life form to fossilize it must first be trapped in mud and buried under conditions where it will not decay. It must then be preserved for very long periods while the atoms that make it up are slowly substituted for by atoms from the rocks, so that the life form, or parts of it, is slowly turned to rock without losing its original shape and appearance. It must then survive geologic vicissitudes long enough to be found by human beings The hard parts of life forms (shells, bones, teeth) fossilize much more easily than the soft parts, so that life-forms without hard parts are rarely found in fossil form.

All in all, the fossil record is not only terribly incomplete but may remain so forever. Still there is enough in it to demonstrate evolutionary change forcibly. It must also be remembered that the scientific view of evolution doesn't depend on fossils alone but on evidence from many branches of science, all of which strongly confirm what the fossils tell us.

The struggle for the acceptance of evolution was no-where so desperate as in the case of the evolution of human beings. It is almost as though people would be ready to accept evolution if only, somehow, an exception could be made in favor of Homo sapiens and if we alone could be allowed to spring ready-made from the mind of God.

Darwin, himself, in *The Origin of Species*, carefully omitted any consideration of human evolution, not because he thought human beings to be exempt from it, but because he didn't want to stir up a storm of controversy. Of course, the book stirred up the storm anyway, and in 1871 Darwin, feeling he had nothing to lose, published *The Descent of Man,* in which he boldly took up human evolution.

And, of course, the storm that resulted was enormous. Since the lower animal that would serve as the human ancestor in the evolutionary view would surely resemble an ape, the question was whether human beings were originally created in the form of apes or of angels. As Benjamin Disraeli (1804–1881), an important British states-man of the time, said (coining a phrase in the process), "I am on the side of the angels."

The point could be argued forever in words alone with-out any settlement. What was needed was some physical evidence of human evolution, and the best and most dra-matic physical evidence would be some fossilized creature that was somewhere between an ape and a human being. (This was widely called "the missing link" in the decades after Darwin's book was published.)

Finding physical evidence was easier said than done. Considering the unlikelihood of fossilization in general, it was very possible that there might be very few examples of early forms of human life that had fossilized. And even if those few did exist, how high was the chance that people might stumble on them, or even, perhaps, recognize them for what they were if they did find them?

To be sure, certain extinct animals were associated with human beings, showing that if a catastrophe were responsible for the wiping out of certain life forms, then human beings must have existed before the catastrophe as well as after.

Thus, in 1799 the carcass of an elephantlike creature was found frozen into a cliffside on the Arctic coast of Siberia. It was not quite a modern elephant, however, for it had a large hump on its skull, a thick coat of long hair, small ears, and unusually long tusks. It was clearly an extinct form of elephant, adapted to a cold climate, and it must have flourished in the Ice Age.

A number of mammoth carcasses were found after that, and in 1860 a French paleontologist Edouard Lartet (1801–1871), discovered in a cave a mammoth tooth that had on it an excellent drawing of a mammoth by someone who, clearly, had seen it in life. The mammoth was hunted by human beings, and perhaps that had contributed to its exinction about 10,000 years ago. There was no question after that that human beings and mammoths coexisted in early times. Again, when the Cro-Magnon skeletons were discovered, they were accompanied by the bones of extinct animals, which the Cro-Magnons had presumably hunted down, killed, and eaten.

This, in itself, would not shake those who supported the biblical account, however. The Bible does describe a catastrophe that was not total—Noah's Flood. Mammoths and other extinct animals associated with human beings might simply not have survived the flood for some reason, and the human beings before the time of Noah might well have hunted them.

Before these discoveries were made, however, and even before Darwin had published his famous book, the discovery was made of skeletons that were clearly human and yet were *not* "modern man."

In western Germany, in the middle course of the Rhine river, is the city of Dusseldorf. Directly to its east along the banks of the small Dussel river is the Neander Valley. The German word for valley is *Tal* or, in more archaic spelling *Thal*. The region east of Dusseldorf is, therefore, *Neandertal,* or *Neanderthal.*

In the Neanderthal, in 1856, workmen were clearing out a limestone cave and came across some bones. This is not an unusual thing to happen, and the logical thing to do is to throw the bones away along with the other debris. This was done, but the word got to a professor at a nearby school. He managed to get to the site and salvage about fourteen of the bones, including a skull.

The bones were clearly human, but the skull, in particular, showed some interesting differences from modern man. It had pronounced bony ridges over the eyes, which ordinary human beings don't have. It also had a backward-sloping forehead, a receding chin, and unusually prominent teeth.

The remains were quickly dubbed *Neanderthal man*, and the question arose as to whether it was a primitive form of human being and the ancestor, perhaps, of modern man. If so, here was human evolution demonstrated.

Naturally, there was strong opposition to such a view. The bones, aside from the skull, were quite human, and the skull itself might merely be that of a deformed human being or of someone suffering from a bone disease. The most prominent scientist to support this view was the anti-evolutionist German biologist Rudolf Virchow (1824–1880).

One very popular suggestion was that the skull was only forty or so years old and was the remains of a Russian soldier who had died during the Russian march into western Europe in 1813 and 1814 in pursuit of Napoleon.

Three years after the discovery, Darwin's book was published and those who were inclined to accept evolution were now eager to interpret Neanderthal man accordingly. In 1863, the English biologist Thomas Henry Huxley (1825–1895), a fierce supporter of Darwin, studied the bones and came out strongly in favor of Neanderthal man being an ancient form of human being ancestral to modern man.

In 1864, another British scientist named Neanderthal man *Homo neanderthalensis*, thus putting it into the same genus as Homo sapiens, but assigning it to a different species.

If the discovery of the bones in the Neanderthal cave had been an isolated incident, the argument might have continued forever. In 1886, however, two similar skeletons were found in a cave in Belgium. The skulls featured all the characteristics of Neanderthal man, and it became very difficult to suggest that all three just happened to have the same abnormal bone disease, one that was never found in modern human beings. The pendulum swung in favor of Homo neanderthalensis as ancestral to Homo sapiens, especially when discoveries of still other such skeletons followed.

Even so, for a half century all one had were scattered bones and remnants of Neanderthal man. It was not till 1908 that the French paleontologist Marcellin Boule (1861–1942) managed to assemble a complete Neanderthal skeleton from a French cave. It was from his reconstruction of how the skeleton must have looked in life that the popular conception arose of Neanderthal man as a short, bandy-legged creature with a repulsive apelike face.

Of course, this was made worse by artists always presenting Neanderthal man as badly needing a shave, while Cro-Magnon man is always shown clean shaven, with a sadly noble expression on his face. (In fact, to those of you who have seen the classic movie *Dr. Jekyll and Mr. Hyde* with Frederic March, Dr. Jekyll was shown precisely as Cro-Magnon man was thought to be, while Mr. Hyde was Neanderthal to the life. I can't believe this was an accident.)

As it happened, though, Boule was working with the badly arthritic and deformed skeleton of an old man. The study of other skeletons of younger individuals in better health that have since appeared makes it seem that Neanderthal was not particularly subhuman. Yes, there are the heavy brow ridges, the large teeth, the protruding mouth region, the receding chin, and the retreating forehead, but on the whole, Neanderthal man stood bolt upright, walked exactly as we do, and showed no important differences from us, from the neck down.

What's more, the Neanderthal brain is as large as ours

and perhaps even a little larger, though it is differently proportioned. The Neanderthal brain is smaller in front (hence the retreating forehead) but larger behind. Since the front part of the brain is associated with the more rarefied regions of abstract thought, we might suppose that the Neanderthals were less intelligent than we—but there is no real evidence of that.

Neanderthal man was apparently shorter than we are, and stockier, with a heavier and stronger musculature, but all the differences do not seem to mean much, biologically. Neanderthal man is now considered to belong to the same species we do, so that the scientific name is now *Homo sapiens neanderthalensis,* while modern man is *Homo sapiens sapiens*.

Neanderthal man lived in Europe for the most part, and more Neanderthal remains have been found in France than anywhere else, but the Neanderthal range seems to have spread eastward as far as central Asia. He appeared first, in typical guise, as far back as 100,000 years ago (though some particularly primitive specimens have been reported up to 250,000 years in age). The Neanderthals became extinct about 35,000 years ago, soon after modern man appeared.

We cannot tell whether modern man appeared somewhere else and invaded Europe, supplanting the Neanderthals, or whether the Neanderthals, changing little by little, produced examples of modern man 40,000 years ago and then, in the space of the next 5,000 years were supplanted by them. The latter seems the more logical.

As to how modern man did the supplanting, whether by war, by intermarriage, or by a mixture of both, we can't tell. The record gives us insufficient guidance.

In any case, Neanderthal man is the earliest example of Homo sapiens we know of, which would make our species at least 100,000 years old and perhaps considerably older.

And yet, if we follow the evolutionary scenario, Neanderthal man couldn't have sprung into existence from nothing. There had to have been still earlier predecessors of human beings that were *not* Homo sapiens, yet resembled human beings more closely than they resembled any other life form, even apes. The name now given to any living organism that more closely resembles a human being than it does an ape is *hominid*.

Modern man is the latest hominid to appear and is the only hominid that now exists, but there must have been earlier and simpler hominids in ancient times. We must now turn, therefore, to a search for the beginnings of hominids.

# 6

## HOMINIDS

The German naturalist Ernst Heinrich Haeckel (1834–1919) was a strong supporter of the idea of biological evolution. He was convinced that early hominids had once existed, even giving them the name *Pithecanthropus,* which is Greek for "ape-man." The term *ape-man* came to be much-used in popular writing, replacing the earlier "missing link."

As the nineteenth century drew to its close, there was a serious search for any fossil traces that might represent such early hominids.

One searcher was a Dutch paleontologist, Marie Eugene Dubois (1858–1940). He reasoned that whereas human beings had spread out all over the world, the apes, far less mobile, had stayed closer to their ancestral regions. Therefore, the apes must have evolved in the places they now inhabit, and the hominids (a variety of ape, after all) must also have evolved there.

As it happens, of the four types of apes, gorillas and chimpanzees live in Africa, while orangutans and gibbons live in southeast Asia and in Indonesia.

Haeckel had speculated that gibbons (the smallest of the apes) were closest to the ancestral form from which all apes descended. Though Haeckel was wrong in this, his notion turned Dubois's eyes toward Indonesia. That land of large islands was then largely controlled by the Dutch and was called the Dutch East Indies. Dubois, as a Dutchman, might have an opportunity to work there.

Things turned out as he wished. He joined the Dutch army, hoping for assignment to the East Indies, and in 1889 he was commissioned by the government to search for fossils in certain Javanese deposits. (Java was the most populous, though not the largest, of the Dutch East Indian islands.)

In Java, Dubois began to search. He had amazingly good fortune. In 1891, near a village named Trinil in south-central Java, he came across some teeth and parts of an ancient skull. The skull showed a retreating forehead and eyebrow ridges, like those of Neanderthal man. The portion of the skull that held the brain was, however, quite small.

The human brain of an adult male weighs about 3.3 pounds (1.5 kilograms), and it has a volume of 88.5 cubic inches (1,450 cubic centimeters). Neanderthal man has a slightly larger brain, with a volume of 91.5 cubic inches (1,500 cubic centimeters). The cavity in the skull located by Dubois had a volume of only 55 cubic inches (900 cubic centimeters). The brain held by such a skull could weigh only about 2 pounds (0.9 kilograms), and it would be only three-fifths the size of an ordinary human brain.

Of course, Dubois might have discovered the skull of a child, but this was apparently not so. When bony ridges develop over the eyes in human beings, they do so in male adults. The eye ridges in women and in children of both sexes are nonexistent. Even in Neanderthals, where the ridges are much more pronounced than in modern human beings, the skulls of the young are comparatively smooth. The skull that Dubois discovered, however, had very pronounced bone ridges and was, therefore, very likely that of an adult.

Still, the brain inside that ancient skull would have been twice as large as the brain of any gorilla now living. The brain was, in other words, intermediate between apes and human beings. The teeth also seemed in some ways to be partway between those of apes and of human beings. Dubois was convinced he had found Haeckel's Pithecanthropus, and that was what he called the skeleton, though most people found it simpler to call it *Java Man*.

Dubois kept on investigating the place where he had discovered the skull and teeth, and in 1892 he found a thighbone only forty-five feet from where he had found the skull. It was at the same level in the rock as the skull had been and seemed as old as the skull, but it looked entirely human. From its shape, it seemed clear that the creature who possessed it originally could stand upright and walk on two legs as easily as a modern human being could.

Dubois was convinced the thigh and the skull had been part of the same individual, so he called Java man *Pithecanthropus erectus* ("the ape-man who stood erect") and published his findings in 1894. This was the first discovery of what was undoubtedly a hominid, with a brain undeniably midway between that of an ape and a human being.

Dubois's report raised an enormous fuss, with anti-evolutionists insisting that Dubois had merely found the head of an idiot. So long as only one such skull was known, there was no way of settling the matter, so Dubois should have labored to find other fossils of the sort. He would not, however. He grew so sick of the yelling and screaming that he locked his bones away for years and wouldn't talk about them any more. The search would have to be conducted by others.

In the late 1930s, another Dutch paleontologist, Gustav von Koenigswald, went to Java and undertook the task. He sought the help of local people. He explained exactly what he was looking for and told them that he would pay ten cents for any piece they brought in, however small. This was a mistake, for anyone who found a bone promptly broke it into small pieces to collect a dime for each piece. Even so, von Koenigswald ended up with three skulls and some pieces of jaw with teeth in place, and in all cases the skulls were small. There might be one human idiot with a small brain, but there wouldn't be four. Java man was truly an early hominid.

Meanwhile, attention turned to China. Chinese doctors thought that if old fossil bones and teeth were ground into powder, they could be used in medicine. For that reason, fossils were to be found in Chinese drugstores. In 1900 one of the old teeth turned out to be rather human in appearance, which inspired a search for human fossils.

About thirty miles southwest of Beijing (once written Peking) is a town called Zhoukoudian (once written Choukoutien), near which there are a number of caves that had been filled with hard earth. They seemed a hopeful place to look for fossils.

In one place in those caves, bits of quartz were found. They should not have been there naturally and might have been brought there by human beings. A Canadian paleontologist, Davidson Black (1884–1934), therefore kept working his way deeper into the cave, inspecting everything.

In 1923, a tooth was found; in 1926 another; in 1927 a third. These teeth were studied carefully: They seemed to be not quite human and not quite ape, either. Black decided they belonged to a hominid that was given the name *Sinanthropus pekinensus* ("Chinaman from Peking"). To the general public, however, it was known as *Peking man*.

In 1929, pieces of a skull, jaw, and teeth were uncovered. After Black's death, the work continued under a German paleontologist, Franz Weidenreich (1873–1948). Eventually, portions of forty different hominids were discovered.

Unfortunately, the Japanese had invaded China and taken over the area in 1937. They allowed the digging to continue, but in 1941 when it looked as though the war might spread and become more serious, the paleontologists decided to send the bones to the United States for safekeeping. Two days after the bones were dispatched, however, the Japanese attacked Pearl Harbor, and in the confusion that followed, the bones were lost and never recovered.

In the time the bones were studied, however, enough was learned to show that Peking man was very much like Java man. Nowadays, paleontologists have decided that Java man and Peking man are both of the same species. What's more, although they are not Homo sapiens, they are close enough to it to be part of the same genus. Therefore, names like Pithecanthropus and Sinanthropus have been done away with. Both are said to be examples of *Homo erectus*.

After World War II, bones of Homo erectus were discovered in Africa and, possibly, in Europe. These hominids, although small brained compared to ourselves, were surprisingly capable. The findings at Zhoukoudian make it seem that it was Homo erectus who first made use of fire about 500,000 years ago.

The Homo erectus near Peking was later in time than that of Java and had a somewhat larger brain. In fact Homo erectus may have first come into existence 1.5 million years ago and have persisted until 250,000 years ago, gradually evolving a larger brain. The brain of Homo erectus might have originally had a volume of 52 cubic inches (850 cubic centimeters), and at the end it may have reached 67 cubic inches (1,100 cubic centimeters).

(Incidentally, lengths of time like 250,000 to 1.5 million years are far too old to be measured by carbon-14 dating methods, or any of the other methods I have mentioned earlier. There are, however, other radioactive breakdowns that are much slower than that of carbon-14, and these very slow breakdowns can be used to measure the age of the rock in which Homo erectus remains are found. I will take up the matter in greater detail later in the book.)

What happened to Homo erectus 250,000 years ago? Most likely, Homo erectus continued to evolve, developing a still larger brain, and became first Homo sapiens neanderthalensis and then Homo sapiens sapiens. There are two or three scraps of bone that seem to come from the period between Homo erectus and Homo sapiens, but not enough to make the connection certain.

Is there any chance that the necessary fossils will be

discovered? Of course! Paleontologists search for them diligently at all times—but the chance is not a good one. All the hominid fossils ever discovered, if put into one heap, would fill a rather small packing crate. Hominids are generally too intelligent to allow themselves to be trapped in mud under conditions where fossilization can take place.

Are there any hominids that are still older than Homo erectus?

In 1931, a British paleontologist, Louis S. B. Leakey (1903–1972), began to dig in the Olduvai Gorge, a place in the East African nation of Tanzania, where sedimentary rock had been laid down for two million years. Leakey thought there might be traces of early hominids in the rock.

In the early 1960s, he discovered three skulls that looked very much like Homo erectus skulls except that the bones were thinner and more delicate and the brains even smaller. The volume of the brain would only have been 49 cubic inches (800 cubic centimeters), and it would have weighed just about half the weight of our own brain.

Leakey called those skulls the remains of *Homo habilis* (''skillful man'') because, small as the brains were, stone tools were found near the bony remains. These small-brained hominids were still intelligent enough to use tools and skillful enough to make them.

Leakey estimated the age of Homo habilis to be about 1.8 million years. It may be that they are very early examples of Homo erectus. It may also be that Homo habilis developed in each of two divergent lines, one toward Homo erectus and one toward Homo sapiens. In that case, Homo erectus came to a dead end. However, it is impossible to tell the exact details without more fossils, and paleontologists, even today, continually argue and speculate about the exact line of descent of modern human beings. What no one argues about is that we descended from primitive hominids, whatever the exact details.

Homo habilis is the oldest hominid to be sfficiently like modern human beings to be placed in the genus Homo; therefore, the genus as a whole might be considered to be 1.8 million years old.

That certainly doesn't mean, however, that Homo habilis is the earliest hominid there is. There may be still simpler, still smaller-brained hominids so different from human beings as to be excluded from the genus Homo, that are nevertheless closer to human beings than they are to apes.

And so there are.

In 1923, an Australian-born physician, Raymond Arthur Dart (b. 1893), went to South Africa to teach at a medical school there. In 1924, he came across a fossil baboon skull on someone's mantelpiece and asked where it came from. It was from a place called Taung, where they were blasting down some limestone cliffs. Dart sent a message to people working at the site, asking for any fossils they might find.

He received a box full of limestone with fossils in it. He isolated the pieces and found that when fitted together, they showed something that looked like the skull of a young ape, except that the hollow for the brain was too large for a young ape. There were no eyebrow ridges. Dart published his observations in 1925 and suggested that the fossil might represent a form of extinct life about halfway between apes and men. He called it *Australopithecus africanus* (Latin for "southern ape from Africa").

At that time, people were still arguing over Dubois's findings in Java and little attention was paid to Dart. In 1934, though, a Scottish paleontologist, Robert Broom (1866–1951), came to South Africa and, thinking that Dart might have come across something important, began to look for more such fossils.

In 1936, he visited limestone caves not far from Johannesburg and found another fossil skull of Australopithecus, an adult one this time. For two years, he kept collecting fossil pieces: a thighbone, another skull, and jaw. These seemed to be somewhat larger creatures than Dart's had been, even allowing for adulthood. Eventually, these were called *Australopithecus robustus,* since they had bones that were thicker and more robust than the earlier specimen.

There are probably a number of different species of these creatures, different enough from ourselves to have a genus of their own, for which the name remains *Australopithecus* even though they are *not* apes. In popular language, they are all lumped together as *australopithecines*.

They are small hominids, some only four feet, even as adults. Their brains are smaller than that of any other hominid that is genus Homo. The brain seems to have a volume of 30 cubic inches (490 cubic centimeters) and may have had a weight of no more than 1.1 pounds. That would be only one-third the weight of our brain and less than the brain weight of a modern gorilla. However, since an australopithecine weighed only one-eighth as much as a gorilla, the australopithecine brain is proportionately much larger.

The australopithecines may have used very simple tools of bone and wood, not having advanced to handling stone, which is apparently restricted to organisms of genus Homo.

In 1977, the American paleontologist Donald Johnson discovered the oldest example of an australopithecine yet found. He discovered enough bones to represent about 40 percent of the entire skeleton, and since they are clearly the remains of a female, the name ''Lucy'' was somehow attached to the skeleton. Its scientific name is *Australopithecus afarensis*. The *afarensis* is derived from the fact that it was located in a section of East Africa called Afars, which is at the southern edge of the Red Sea.

Lucy, apparently a young adult, is only about three and a half feet tall. Her hipbones and thighbones confirmed something that had already been suspected from the other australopithecine fossils: She walked fully upright and just as easily as we do.

The hominids, all of them, right down to the earliest we know of, had a unique, double-curved spine that could support them upright indefinitely. Apes, although they can walk upright, do so only for short times and clearly find the process uncomfortable.

It would seem then that the evolutionary development that made hominids, and eventually human beings, possible, was not a giant brain, or a clever hand, but rather a twist to the spine that made it possible to stand upright. From this, all else may have followed.

Once a hominid stood upright, its forelimbs were completely freed from the task of body support. The forelimbs then were freed for manipulating and inspecting surrounding objects. Any change that made the hands and eyes more suitable for this purpose improved the ability of the organism to survive. It meant longer life and more young to inherit the better and more nimble hands, the longer and opposable thumbs, and the sharper eyes.

The more the hands and eyes were used to handle and to inspect, the more information flooded into the brain. And again, any change that happened to make the brain larger and more complex was therefore useful and encouraged survival. This, too, meant longer lives and more young, who inherited the better brains—which have tripled in size during the time lapse from the australopithecine to the present.

Lucy is about 4 million years old. She may not be the oldest australopithecine, nor the first living thing capable of standing upright and of walking freely on two legs, but she is the oldest of which we know. Some paleontologists believe that the australopithecines may have gotten their start a couple of million years earlier still with an original brain volume of only 21 cubic inches (350 cubic centimeters) and a brain weight of only 0.8 pounds, but we'll need more and older fossils before we can really know.

In a way, though, we have not located the "missing link." Even Lucy, the oldest known australopithecine, is much closer to the human being than to the ape because of her ability to walk upright. She is not the "ape-man," not the living organism halfway between apes and human beings that people have searched for.

There are two possibilities that raised hopes in that direction, but both proved to be false alarms.

In 1935, von Koenigswald (who was soon to go to Java to search for more fossils of Homo erectus) came across four interesting teeth in Hong Kong drugstores. They looked just like human teeth, but they were much larger.

Up to that time (and even since, in fact) all the early hominids have turned out to be smaller than Homo sapiens. Even the Neanderthal variety of Homo sapiens, which seems to have been stronger and more muscular than modern human beings, was not as tall as we are. In a way, Homo sapiens sapiens is the hominid giant.

The teeth that Koenigswald uncovered, however, if they were of hominid origin, would have to belong to hominids considerably larger than ourselves. Von Koenigswald didn't quite dare suppose this to be so. He labeled the creature to which the teeth belonged *Gigantopithecus* (Greek for "giant ape").

Of course, people were ready to believe that giant hominids might once have existed. The Bible itself, in Genesis 6:4, says in an oft-quoted line, "There were giants in the Earth in those days . . ." The Hebrew word *nephillim*, translated as "giants" in this verse may not mean giants in the sense of size alone, however. It may simply mean heroic men, great warriors, semi-divine legendary heroes. Still, most people who accept the Bible literally do take the word to mean people of great size.

Then, too, in folktales of many nations there are stories of giants, hominids of great mass and stature, but usually stupid and easy to fool. Are these stories a distant memory of ape-men, or are they just the storyteller's usual way of magnifying difficulties and villains to make the hero seem more heroic? Is it just the David and Goliath situation, with everyone rooting for little David?

In 1955, Chinese scientists decided to poke through all the drugstores they could, in order to find any further parts of the creature that might exist. They discovered dozens of giant teeth and a couple of giant lower jaws.

It turned out that Gigantopithecus was exactly what the name meant. It was not a hominid at all, but a giant ape about nine feet tall, the largest ape that ever lived so far as we know (though it was far short of that famous and beloved monster, King Kong). It had human-looking teeth because it was adapted for the same sort of diet that human beings had, but its jawbones were unmistakeably apelike.

Gigantopithecus may not have become extinct until the time of early Neanderthal man, so it is conceivable that it may have helped give rise to the legend of stupid giants, but somehow I doubt it.

Even more puzzling was the case of a find made in 1911 in Piltdown, in southern England, by an English lawyer, Charles Dawson (1864–1916). It consisted of a skull and, later, of a lower jaw with some teeth. The skull seemed quite human, but the jaw seemed quite apelike. It was named *Eoanthropus dawsoni* (Greek for "Dawson's dawn man"), and it was commonly called Piltdown man.

With its human skull and apelike jaw, could it be the halfway apeman, the missing link?

For forty years, it puzzled paleontologists. In all other hominids, as the skull grew more human, the jaw grew more human, too. A hominid with a human skull and an apelike jaw just didn't seem right. As more and more fossils were discovered, Piltdown man seemed less and less right, but the paleontologists that had first fitted the skull and jawbone together defended it bitterly.

Well, it *wasn't* right. By 1953, it was clearly proved that Piltdown man was a fake. The skull *was* human, and quite recent. The jaw was that of an orangutan, also recent. The bones had all been treated to make them look very old, the teeth had been filed down to make them fit. The connections between jaw and skull had been broken away so that one couldn't see that they clearly *didn't* fit each other.

The key proof that both parts were modern was fluorine analysis. Bone, as it exists in the body, contains few or no atoms of an element called *fluorine*. As bone lies in the ground under fossilizing conditions, however, it very slowly absorbs fluorine from the soil and from the water in the soil. From the amount of fluorine in the fossil one can get a rough idea of how long it has lingered in the soil.

Who could have carried out such a hoax? Most people suspect Dawson, but the matter can't be proved and half a dozen other people are also suspected. Nor has anyone figured out the motive. It remains the most famous hoax—an unsolved hoax at that—in science.

Of course, the fake was easy to see after it was revealed, and there is considerable wonder as to how so many learned professors could have fallen for it.

In part, the reason was that in 1911 very little was known about early hominids. Nowadays, anyone attempting to foist the combination of a human skull and an ape jaw on paleontologists would be kicked out at once, for paleontologists know enough now to know that this combination is extremely unlikely. But they didn't then.

Then, too, paleontologists are human, and it was a matter of national pride. Although fossil finds had been made in Spain, France, Germany, and Belgium, very little in the way of hominid relics were found in England. When the chance came for English paleontologists to lord it over the rest of the continent with a relic so unprecedented and unusual, they simply couldn't resist. They called it "the first Englishman" and insisted on its authenticity.

But even if we haven't found the true link between hominids and apes, we can be sure that the first hominid didn't arise out of nothing. Human beings and apes are sometimes lumped together as *hominoids,* and there must have been a first hominoid, some creature from which all the apes (and human beings too) descended, and which earlier had been split off from the monkeys.

If you add the monkeys as well and some still more primitive creatures, you have an *order* called *Primate,* from a Latin word meaning "first." Our next step, then, is to investigate the beginnings of hominoids and primates.

# 7

## PRIMATES

Already we have moved far back in time, much farther than anyone would have dreamed possible two centuries ago. If we consider the hominid line to be 6 million years old, then for three-quarters of that period australopithecines were the only hominids alive. Only in the final quarter of hominid history did genus Homo appear, and 98 percent of it was over before Homo sapiens neanderthalensis appeared. About 99.3 percent was over before Homo sapiens sapiens appeared, and we have been civilized for only 1/600th of the time that hominids have existed.

Yet it is clear the evolutionary history of hominids extends far back beyond their first appearance.

It is not necessary to be an evolutionist to realize that apes and monkeys look like us. Even the ancients appreciated the fact that monkeys are almost caricatures of human beings. In fact, although the word *monkey* is of uncertain origin, I like to think it assumed the form it has, in English, because of the resemblance of the sound to that of *manikin*.

The ancient inhabitants of the Mediterranean were only

familiar with the monkey branch of the Primate order (excluding human beings themselves, of course) but the human resemblance was unmistakable. Their faces were those of shriveled little men. They had hands that clearly resembled human hands, and they fingered things as human beings did, doing so with a lively curiosity. They were visibly more intelligent than other animals.

However, they had tails and that rather saved the day. The human being is so pronouncedly tailless and most of the animals we know are so pronouncedly tailed that that difference, almost by itself, would seem to indicate the uniqueness of human beings and put us in a class by ourselves.

There is a reference to a monkey in the Bible, however, for which the translator used a special word. In discussing King Solomon's trading ventures, the Bible says in 1 Kings 10:22, ". . . once in three years came the navy of Tharshish, bringing gold, and silver, ivory, and apes, and peacocks."

Tharshish is usually identified with Tartessus, a city on the Spanish coast just west of the Strait of Gibraltar. In northwestern Africa, across from Tartessus, there existed then (and now) a type of monkey of the macaque group. It was this macaque that was called an "ape." In later years, when northwestern Africa became part of Barbary (because it fell under the control of "barbarians") it was called "Barbary ape." Some of these apes exist on the British-owned Spanish peninsula of Gibraltar and are the only monkeys native to Europe.

The odd thing about the Barbary ape, and the characteristic that makes it seem to deserve a special name of "ape" rather than "monkey," is that it lacks a tail. It therefore resembles human beings more than other monkeys do. When the Greek philosopher Aristotle (384–322 B.C.) prepared his classification of life forms, he placed the Barbary ape at the top of the monkey group, just under man, entirely because of its taillessness.

The Greek physician Galen (130–200) was not satisfied to go by the superficial appearance. He dissected Barbary apes and reported that the muscles, bones, and internal organs all bore an uncanny resemblance to those of men.

In medieval times, many people resented this resemblance. Told by the Bible that human beings had been made in the image of God (and accepting that phrase literally rather than symbolically), they didn't want mere animals intruding on that image. There was a tendency to look at monkeys as somehow in league with the devil, and perhaps made in the devil's image as human beings were made in God's.

Monkeys, however, weren't the worst of it. There were other creatures, unknown to Europeans of ancient and medieval times, that were larger than monkeys and that resembled human beings even more closely. They were like the Barbary ape in being tailless, so they too were considered apes. Because of their particular resemblance to human beings, they were distinguished from the Barbary ape in being called *anthropoid* (Greek for "manlike") apes.

In 1641 a description was published of an animal brought from Africa and kept in the Netherlands in a menagerie belonging to the prince of Orange. From the description it seems to have been a chimpanzee. There were also reports of a large manlike animal in Borneo, one we now call *orangutan*. (Orangutan means "man of the wild" in Malay, and such is its likeness to human beings that some natives were convinced it could talk but didn't do so because it feared it would be put to work if it did). The two other anthropoid apes, the gorilla and the several species of gibbons, were discovered later. The gibbons are the smallest of the anthropoid apes, and the other three— gorilla, chimpanzee, and orangutan—are sometimes lumped together as the "great apes."

Linnaeus, in forming the order which he named Primate, knew enough about the anthropoid apes to find himself forced to include Homo sapiens in the order even though he fully accepted the biblical account of creation. From what he heard of the orangutan, he overestimated its manlike nature and included it in genus Homo, as *Homo troglodytes* (cave-dwelling man). This was wrong, of course.

Of living primates, the gorilla is the largest. A male gorilla is about the height of a man, and his weight can reach 400 pounds. (The female is considerably smaller.) The gorilla is the only primate larger than man, and only the extinct primate Gigantopithecus was larger still.

If we are going to consider the beginning of the hominoids and the primates, it would be best to get an idea as to how the Earth's history has been divided in terms of fossils.

Those portions of Earth's history that are characterized by plentiful fossil remains in the sedimentary rock layers are divided into three major divisions, or eras. These are the *Paleozoic* (Greek for "old life"), the *Mesozoic* ("middle life"), and the *Cenozoic* ("recent life").

As the name implies, the Paleozoic includes the oldest strata and usually the ones most deeply buried. The Cenozoic are the most recent strata, which are also topmost, and the Mesozoic lies in between. The dividing lines come at places where there is a more or less sudden change in the nature of the fossils present.

For the moment we will be concerned with the Cenozoic, the most recent one, which covers the last 65 million years of Earth's history.

The Cenozoic is divided into seven subdivisions or *epochs*. The time of each is given in the following table as so many "million years ago" (MYA).

Paleocene ("old-recent"), 65–54 MYA
Eocene ("dawn of the recent"), 54–38 MYA
Oligocene ("a little of the recent"), 38–26 MYA
Miocene ("less of the recent"), 26–7 MYA
Pliocene ("more of the recent"), 7–2.5 MYA
Pleistocene ("most of the recent"), 2.5–0.01 MYA
Holocene ("entirely recent"), the last 10,000 years

The Holocene, which is the most recent epoch, and the one in which we are living now, includes all of civilization, from the invention of agriculture.

The Pleistocene includes the entire history of genus Homo.'

The Pliocene includes the entire history of the australopithecines.

To investigate the beginnings of the hominoids and the primates, generally, we must move back beyond the Pliocene.

In 1934 an American paleontologist, G. Edward Lewis, came across some teeth and pieces of jaw in ancient deposits in the Siwalik Hill in northern India. They were in rocks that were too ancient even for australopithecines. The fossils were over 7 million years old and therefore had to come from late in the Miocene.

Lewis wasn't sure whether these fossils represented a hominid or not. If it was a hominid it was even earlier and more primitive than the australopithecines, but that was a very difficult decision to make from teeth alone. He called the fossil *Ramapithecus* or "ape of Rama," Rama being one of the important Hindu gods of India. Very similar remains are assigned to *Sivapithecus,* Siva being another Hindu god.

Primate fossils that resemble apes more than they resemble human beings are called *pongids,* and it may be that Ramapithecus is close to the borderline between hominids and pongids and may fall either way. What is desperately needed are thigh and hipbones to see if Ramapithecus walked upright or not. At the moment, paleontologists lean toward the pongid and suspect that Ramapithecus walked gorilla fashion rather than human fashion. It is also suspected that Ramapithecus and Sivapithecus first evolved 14 million years ago.

Louis Leakey and his wife, Mary, digging along the shores of Lake Victoria in east Africa, came across the bones of what was clearly an extinct ape. There was no question about that, for its jaws and teeth were very apelike.

Leakey named it in honor of a chimpanzee in the London Zoo who was called Consul and who was a great favorite with the public. Leakey called the new find *Proconsul* meaning "before Consul." Eventually, a number of Proconsul bones were found, including an almost complete skeleton, so paleontologists could see in what ways it was more primitive than present-day apes.

Proconsul seems to be a member of a group of species of primitive apes, all of which belong to a genus called *Dryopithecus* ("oak-tree apes") because the fossils were found along with some traces of ancient oak forests.

There were apparently Dryopithecus species of different sizes, some no larger than a smallish monkey, but some almost as large as a gorilla. The earliest species seem to have developed about 25 million years ago, just at the beginning of the Miocene.

Dryopithecus seems to be the common ancestor of present-day chimpanzees and gorillas, but the question is whether it also gave rise to Ramapithecus and to the hominids. We can't answer that question yet, but certainly Dryopithecus seems to be a possible candidate for the common ancestor of the great apes and human beings.

At about the same time as Dryopithecus, there are fossil remains assigned to *Pliopithecus,* which may possibly be the ancestor of the gibbons, the smallest of the anthropoid apes.

If we move back to the Oligocene, there are some scraps of fossils that have been given the name *Aegyptopithecus* ("Egyptian ape") because they were located in Egypt. Aegyptopithecus may have evolved as much as 40 million years ago in the late Eocene. It, or something like it, may represent the general ancestor of the hominoids—all the pongids and hominids.

We must go farther back into the Eocene and the Paleocene to go back to the fossils of very primitive primates that gave rise to the entire order, including not only the hominoids, but all the species of monkeys together with groups of animals even more primitive than monkeys that are still members of the Primate order.

More primitive than the monkeys, for instance, are the lemurs, which are today represented most commonly on the island of Madagascar, off the coast of southeast Africa. They are more squirrellike than monkeylike in appearance but have sufficient resemblance to monkeys to be placed in the Primate order. Some 50 million years ago in the early Eocene, the lemurs were flourishing and from them the monkeys and apes evolved.

Even more primitive than the lemurs are the tree shrews, which are only hesitantly classified as primates by some taxonomists. They seem to have as much or more in common with insectivores, like the shrews and hedgehogs. The earliest primate to evolve may well have been tree-shrewish in appearance. Some teeth have been located in the early Paleocene, about 60 million years ago, for a creature about the size of a rat. It has been named *Purgatorius,* and this (it is just possible) may have been close to the ancestral primate.

But earlier than the Cenozoic, we have the Mesozoic, and we can trace the evolutionary process into this still-earlier time if we are willing to move from the Primate to a wider group, the class *Mammalia.* That is what we'll take up next.

# 8

## MAMMALS

The Primate order is one of twenty orders, all of which belong to the class Mammalia. All species of Mammalia ("mammals") share certain characteristics. All mammals have hair; all have a diaphragm; all but a very few give birth to living young, usually with the aid of a placenta; the young are all fed on milk, produced by the mother and delivered, with very few exceptions, from breasts. (It is the breasts or, in Latin, *mammae* that give the class its name.)

Included among the mammals are (just as a sampling): anteaters, hedgehogs, bats, rabbits, rats, seals, whales, cats, dogs, bears, elephants, horses, cattle, sheep, goats, monkeys, and, of course, human beings.

These are actually a varied lot. Most of them are land animals, but whales and dolphins live permanently in water, while bats are as much at home in the air as birds are. The largest mammal, the blue whale, can be 100 feet long, with a weight of up to 150 tons. It is not only the largest mammal, but the largest animal of any kind, not only now, but ever. If you're thinking of dinosaurs, the blue whale is twice as heavy as the heaviest dinosaur that ever lived.

The smallest mammals are at a serious disadvantage, because mammals are warm-blooded and must keep their temperatures quite high. (The normal human body temperature is 98.6 Fahrenheit). The smaller the mammal, the larger its surface compared to its weight and the more rapidly it loses the heat it can generate. The smallest mammals are tiny shrews only a couple of inches long, including the tail, and weighing only one-fifteenth of an ounce. They must be eating almost every waking moment in order to keep stoking the metabolic processes.

We think of the mammals as the rulers of the Earth, and they are certainly the most intelligent animals. However, they are not doing well.

Human beings, to be sure, are doing well. In the course of the Holocene epoch, the 10,000 years of civilization, the human population has increased from 4 million to 5,000 million, a 1,250-fold increase. The domestic animals that people protect and make use of have also vastly increased in numbers.

However, the Earth in general can only support so much animal life, and for every additional pound of human beings and his animal favorites, a pound of other animal life must disappear. It is not surprising, then, that during the Holocene some large mammals have become extinct. These include the mammoth and the mastodon, each a variety of elephant; the ground sloth of South America; the Irish elk, with the largest antlers of any deer that has ever lived; the cave bear; the aurochs, which was the wild ancestor of cattle; and so on.

There is some argument as to whether they were hunted to death by human beings, or whether their extinction was the result of some climatic change.

To my mind (as a nonexpert), it seems a silly thing to argue over. Of course, human beings were reponsible. Even if human beings didn't actively hunt the animals to death, which I bet they did, they gradually took up the living space. Large mammals are particularly vulnerable under such conditions. They require a great deal of food and, therefore, a great deal of space within which to find their food. They are relatively few in number, at best. They grow slowly, and have few young and those at comparatively long intervals. An unusual number of deaths among large mammals is, therefore, a much greater drain on them as a species as the same number of deaths among smaller, more fecund species.

Even those large mammals who have not yet been driven to extinction, and whom humanity is belatedly trying to protect, are nevertheless in a bad way. Their living space is much reduced, and they are in danger of extinction in the near future.

All this, however, does not mean that the end of mammals is necessarily here. The small mammals are still holding their own. Consider the rat, against whom the hand of humanity is mercilessly turned. The rat manages well, living in the crannies of our living spaces, feeding on what he can steal of our food, and breeding new rats as fast as old rats are killed.

In the Pliocene, on the other hand, when the australopithecines were making their appearance and when the hominids were not yet an important factor, large mammals filled the globe. And earlier, in the Eocene and the Oligocene, there was a kind of golden age of large mammals. The *Titanotheres* ("titanic beasts") flourished then, between 50 and 35 million years ago. They are large, hooved herbivores with small brains and, frequently, grotesque horns on their heads. They can't be viewed as failures since they lasted at least 15 million years, but they did become extinct in the middle of the Oligocene, between 30 and 40 million years ago.

This is one of those "mass extinctions" that take place on Earth now and then; sometimes very drastic ones. Paleontologists are vigorously arguing about the matter, seeking causes, and I will discuss the matter in some detail later in the book. As for the Oligocene extinction, that might have taken place because tough grasses were spreading, and it is possible that the titanotheres didn't have and, for some reason, didn't happen to develop the kind of teeth needed to eat them. Or else they fell prey to carnivores, who were becoming brainier and against whom the stupid titans had no adequate defense. Or else there might have been, as we shall see, a more dramatic catastrophe.

The largest of all the land mammals who ever lived was the *Baluchitherium* ("beast from Baluchistan"). Its fossil remnants were discovered in Baluchistan (in what is now Pakistan) in 1907 by the American zoologist Roy Chapman Andrews (1884–1960).

The Baluchitherium was a large, hornless rhinoceros that stood 18 feet (5.4 meters) tall at the shoulder, so that these shoulders were as high off the ground as the head of a tall giraffe. The Baluchitherium's head, when he stretched it upward, could reach 26 feet off the ground. His weight might be as high as 30 tons, or three times that of the largest African elephant that ever lived.

Why did the mammals become so large in the Eocene and Oligocene? They were much smaller before that and became rather smaller after that. The answer is no puzzle.

For a long period of time before the Cenozoic era (sometimes called the "age of mammals"), giant reptiles dominated the land areas of the globe. Some of them were even larger than the largest mammals the later Cenozoic ever produced, and while these reptiles existed, the mammals couldn't evolve large size. They'd be invading the environmental niches occupied by the reptiles and would be killed by them. The only successful way of mammalian survival was to be small and fecund, to be nocturnal, to live in burrows. In short, the only way mammals could survive was to make sure they went largely unnoticed by the ruling reptiles.

However, about 65 million years ago, the large reptiles and many other kinds of organisms died off in one of the really large mass extinctions.

Whatever the reasons for this "great dying," as it is sometimes called, the effect was that of leaving a massive environmental niche open. If a mammal happened to increase in size, there were no giant reptiles from whom to

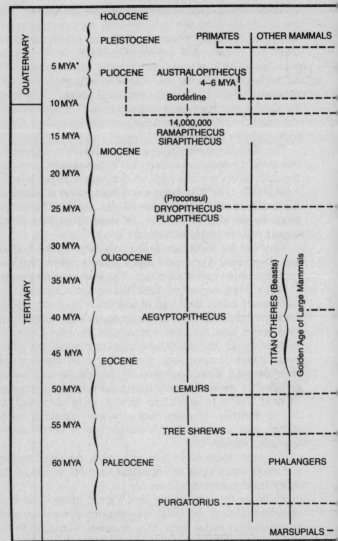

# AGE OF MAMMALS
(Cenozoic)

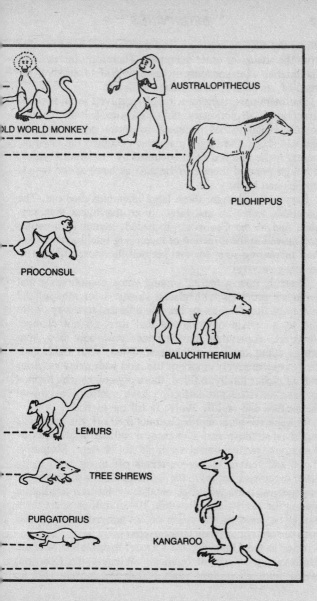

attract unwelcome attention, and the mammal became safer from the attacks of other mammals. Therefore, increases in mammalian size suddenly became an aid to survival instead of, as before, an invitation to death.

The mammals, therefore, rapidly evolved in all sorts of directions ("evolutionary radiation") to fill the various environmental niches occupied by those organisms who had vanished. The larger mammals occupied the environmental niche that had been occupied by the larger reptiles, although none of them ever became as large as the largest reptiles had been.

Eventually, though, these large mammals died out. The mammals were, by and large, more intelligent than reptiles, and as the Cenozoic proceeded, mammalian evolution moved in the direction of increasing intelligence rather than increasing size, for that seemed the more efficient in ensuring survival.

There is now a feeling among some evolutionists that evolution proceeds with glacial slowness for most of the history of life. Organisms become adjusted to a way of life and to a certain environment and then do not change. However, something may happen, now and then, that brings about massive extinctions. Then, with Earth suddenly comparatively empty of life, and with many environmental niches totally unfilled, those organisms who happen to have survived the extinction have a chance to spread themselves and rapidly evolve to fill the niches.

Thus, if the giant reptiles had not died off, the mammals might never have had a chance to radiate in all sorts of directions, and we ourselves might not be here. Similarly, if we succeed in killing ourselves off along with many other organisms but leave the Earth viable for some surviving species, there would be another evolutionary radiation among the survivors and, within 10–20 million years there would be wide variety again on a totally different basis, with absolutely unpredictable results.

Prior to the Cenozoic, the "age of mammals," was the Mesozoic, the "age of reptiles." Whereas the Cenozoic lasted from 65 million years before the present (MYA, remember) to the present for a total of 65 million years, the Mesozoic lasted from 225 MYA to 65 MYA, for a total duration of 160 million years. The Mesozoic, in other words, lasted about two and a half times as long as the Cenozoic but, of course, the Cenozoic is still going on.

The Mesozoic is divided into three periods. Of these, the latest is the Cretaceous, from a Latin word meaning "chalky," because chalk is characteristic of many rocks laid down in that period—like the famous white Cliffs of Dover, for instance. It endured from 135 MYA to 65 MYA for a total of 70 million years. The Cretaceous is, in itself, longer than the entire Cenozoic era.

Before the Cretaceous is the Jurassic, named for the Jura mountains on the border of France and Switzerland, where the first rocks assigned to this period were studied. It extends from 190 MYA to 135 MYA for a duration of 55 million years.

Finally, the oldest part of the Mesozoic is the Triassic, from the Latin word for "three" because the rocks that were first studied from this period consisted of three layers. It extended from 225 MYA to 190 MYA for a duration of 35 million years.

If we trace the mammals back into the Cretaceous, there is no sign of the monsters that are to appear later. They are just small creatures, obscure and apparently unimportant, and among them are those that will eventually give rise to the first primates.

All the mammals I have mentioned so far are placental mammals or *Eutheria* (from Greek words meaning "true beasts"). These are the dominant form of mammals and have been so throughout the Cenozoic. The placental mammals bring forth living young with the help of a placenta, a complex organ that makes it possible for food to diffuse from the mother's bloodstream into the fetus's bloodstream and for wastes to diffuse in the opposite direction. (There is no direct connection of bloodstreams, however.)

This enables the fetus to remain in the mother's body for a long time (nine months in the case of a human being, two years in the case of an elephant) and to be born relatively advanced.

Placental mammals came into existence toward the end of the Cretaceous and were small organisms that probably lived on a diet of insects.

There are, however, mammals that are not placental and that have a simpler reproductive system. The young are born alive, but very prematurely by placental standards, and must crawl from the mother's vagina to a pouch on her abdomen. Inside the pouch are the nipples where the young (actually embryos) feed on milk until they are capable of independent life. Such mammals are called *marsupials*, from the Latin word for "pouch."

The marsupials developed at about the same time as the placentals some 75 to 80 million years ago, toward the end of the Cretaceous. In the evolutionary radiation of the mammals after the great reptiles had disappeared, the marsupials also produced large beasts, some as large as elephants. The marsupials for the most part evolved in the southern part of the land masses of those times, and the placentals in the northern part.

On the whole, though, the marsupials did not compete well with placentals when the two lived in the same regions. As the placental mammals made their way southward, the marsupials died out.

Marsupials remained dominant only in Australia and some nearby islands, but that seems to have been only because placental animals that presumably existed in Asia could not cross the stretches of water into the Australian regions. Bats could, of course, and eventually man did, bringing the dog with him. Then when European settlers arrived in Australia toward the end of the eighteenth century, they brought other placental mammals with them and the marsupial population is now declining even in Australia.

The largest and best known of the marsupials still alive is the red kangaroo, which can be the size and weight of a man. In the American continents there are various small opossums, the only marsupials living outside the Australian regions. They flourish despite placental competition, in part because they are so fecund.

Ancestral to both the placentals and the marsupials are a group called the *Pantotheria* (from Greek words meaning "all beasts," because virtually all the mammals may have descended from them). Their fossil traces are found in the Jurassic, perhaps 150 million years ago, and the best example found is a small hopping animal with an apparently primitive marsupial-type system of reproduction. Marsupialism would therefore seem to be, as is not surprising, older than placentalism.

There were still older and less advanced mammals, of which there are some survivors even today. These include the duckbill platypus and the echidna, native to Australia and New Guinea. They have hair and produce milk, so they are certainly mammals, but they are only imperfectly warm-blooded, their inner temperature varying more widely than those of other mammals.

The most amazing thing about these mammals, though, is that they lay eggs very much like those laid by reptiles. (European biologists flatly refused to believe this when the news first reached them). The skeletons of these mammals possess certain reptilian characteristics also.

These mammals are called *monotremes* (from Latin words meaning "one hole"), since instead of possessing one opening for feces, a second for urine and, in the case of females, a third for the delivery of young, (as is true for all other mammals), these have but one opening, as in reptiles and birds, for feces, urine, and egg-laying.

The very first, most primitive mammals, only the size of mice and shrews, and surely egg-laying, appeared in the Triassic, perhaps 200 million years ago. For the first two-thirds of their existence, then, mammals were such insignificant creatures that zoologists (if any had existed in the Mesozoic) would scarcely have wasted a footnote upon them.

Naturally even the proto-mammalian "mice" of the Triassic had to come from somewhere, but before we trace them further back, let's point out that mammals are not the only warm-blooded animals. There is another group, the birds, that are, on the whole, even warmer than mammals are to a slight degree. Since what is most notable about the birds (at least to our envious eyes) is the ability to fly, and since I have taken up the beginnings of human flight at the start of the book, let's consider, next, the beginnings of animal flight.

# 9

## ANIMAL FLIGHT

Animals have, on four different occasions, developed the ability to fly through the atmosphere. Each time their bodies adapted for the purpose in slightly different ways.

The most recent development of animal flight has been by the bats, the one group of mammals capable of true flight. The bats, like mammals generally, have hair, bear live young with the aid of a placenta, and suckle their young on milk. Their forelegs have long finger bones along which are stretched a thin membrane that often stretches backward to include the leg bones also. The feet are free and can be used by the bat to crawl about (with the aid of the folded wings acting as clumsy arms) when necessary. The bat can also suspend itself from a branch by means of its feet. The clawed thumb on each hand also remains free. The order to which the bats belong is called *Chiroptera* (Greek for "hand wings," for obvious reasons).

The bats are a successful group of animals, and make up 900 species spread all over the world, thanks to their ability to fly. The smaller ones eat insects, the larger ones fruit, and they tend to be nocturnal. It is not their eyes they use for catching insects in the dark, but their ears. They emit short, sharp squeaks, mostly *ultrasonic*, that is, too high-pitched for human ears, and pick up the echoes. From the direction from which the echo comes and from the time it takes it to return, they can detect an insect, or an obstacle, and its position as well as we can with our eyes.

During World War I, a French physicist, Paul Langevin (1872–1946), worked on a device for detecting submarines by beams of ultrasonic sound. It was eventually perfected and was called *sonar*, or *echolocation*. Bats, however, had that precise system worked out very nicely millions of years before we did.

Bats are small creatures. The largest known bat is a fruit eater from Indonesia. It can be nearly 16 inches (40 centimeters) from nose to tail, and has a wingspread of nearly 6 feet (1.8 meters). It is mostly membrane, however, and its total weight doesn't quite reach 2 pounds (0.9 kilograms). The smallest species of bat weighs less than an ounce.

This is not surprising. Air is not a very buoyant medium. One must expose a large surface area of wing in order to get enough lifting property and use a considerable muscular effort in order to forge one's way upward by the beating of those wings. As the body size grows larger, the mass increases quickly and the wings have to become longer and longer in proportion. At some not very large weight, flying by the use of the body's own muscles simply becomes impossible.

It used to be thought that human muscles, for instance, were simply not strong enough to maintain the human body in the air, regardless of what wing surfaces were attached. Recently, however, a very light glider with very efficient wings was forced through the air across the narrowest part of the English Channel by the use of bicycle pedals turning a propeller. However, the device just barely lifted above the water surface, just barely made it across the sea, and was more a demonstration that it could be done then evidence of anything practical or even useful.

A flying horse kept aloft by wings under muscular power alone is, of course, unthinkable, and Pegasus only inhabits legend. The fact that heavy airplanes, weighing many tons, can fly easily rests on the fact that they are not run by muscles, but by engines that deliver far more power than muscles can.

So far, paleontologists have not been able to trace the beginnings of bat flight. The oldest fossils that are clearly bats are about 45 million years old and are from the Eocene, but at that time the wings were already fully developed and we still don't have evidence of what the earlier stages were.

We can assume that there must have been a preliminary period in which membranes were being developed but could only be used for gliding. There are, after all, mammals that glide. The flying squirrel is about the best known of these. It can stretch out all four legs and its loose membranous skin converts the animal into a living kite. They can glide long distances but it is not true flight for they cannot gain altitude at will.

These are also lemurs (primitive primates) and phalangers (which are marsupials) that can glide in the same way. There are lizards that can glide on expanded, membranous feet, and "flying" fish that can glide through the air on enlarged fins.

Putting all this aside, it is the birds that are the flyers par excellence. Since they are flyers, they are, on the whole, small creatures and can easily lose body heat. Since flying is a most energetic activity, they must maintain a body temperature slightly higher than that of mammals.

In order to maintain a high temperature against the tendency to lose heat, birds must conserve heat and for this they have feathers. Feathers are a more efficient insulating device than hair is and they are formed only by birds. No organism that is not a bird has ever produced a feather, so far as we know, and no birds are completely without them.

The bones in bird wings, unlike those in bat wings, are fused together. It is the strong, long wing feathers in birds that present a surface to the air and make flying possible, not membranes, as in bats.

Like mammals, birds evolved in the Mesozoic at a time

when reptiles were dominant. In some respects, birds are closer to reptiles than mammals are. Birds have not specialized in the large brain that mammals have produced. They lay eggs as reptiles do, and their skeletons are more reptilian than the mammalian skeleton is.

The largest birds capable of flight probably do not weigh more than 40 pounds (18 kilograms). Still, this is twenty times the weight of the largest bat and speaks for the powerful flying muscles birds have and the efficiency of their flight-machinery. Some albatrosses, which are among the heavier flying birds, have a wingspread of up to 10 feet (3 meters).

The smallest bird is the bee hummingbird, which weighs less than a tenth of an ounce (2 grams) and is smaller than several kinds of large insects. The bee hummingbird is as small as the smallest shrew and this, apparently, is as small as any warm-blooded organism can be—and, even so, the bee hummingbird requires constant feeding.

When the great reptiles dominated the land, the birds, which could fly and thus escape the reptilian jaws, were more secure than the early mammals were. They could grow larger. Once the great dying at the end of the Cretaceous finished off the large reptiles, there was a tendency among the birds, as well as among the mammals, to grow enormous and fill the reptilian environmental niche.

Really large birds could not fly, of course, and they had no need of powerful wing muscles. In flying birds, the breastbone has a keel to which the wing muscles are firmly attached. In large, nonflying birds, such a keel is absent and the breastbone is as flat as a raft. Such large birds are therefore called *ratites* (from a Latin word for "raft").

The ratites flourished particularly on islands. For one thing, there was a tendency for island birds to lose the ability of flight, since attempting to fly on islands entailed the constant danger of being blown out to sea. Secondly, the small birds from whom the ratites descended could reach islands by flying there, while mammals, generally, could not reach them. There would be a period of time, then, when large birds could evolve without the competition of the still more menacing large mammals.

The largest ratite still alive in the ostrich, which, in some cases, can stand 9 feet (2.75 meters) tall and weigh nearly 300 pounds (135 kilograms). Even taller, however, was the giant moa of New Zealand, an island on which no mammals except bats existed until they were brought in by human beings. The giant moa, which was hunted to extinction by the native Maoris during the 1600s, had a long neck that reached to a height of up to 13 feet (4 meters) and a weight of about 500 pounds (225 kilograms).

Still heavier is the *Aepyornis* (from Greek words meaning "tall bird") of Madagascar. It stood only 10 feet (3 meters) tall, but a large specimen may have weighed up to 1,000 pounds (450 kilograms). It may have survived into historic times, for some people think it was the inspiration for the *roc*, the gigantic flying bird in the Sinbad tales of *The Arabian Nights*.

If we move back in time, the earliest fossil we have of a bird with a keeled breastbone is the *Ichthyornis* (Greek for "fish· bird," because it was thought to live on fish). It dates from the late Cretaceous, about 70 million years ago, and has interesting reptilian characteristics. For instance, it had small teeth in its beak, whereas modern birds are, one and all, toothless.

Before Ichthyornis, birds presumably lacked the keel and had relatively weak flight muscles; they could flutter out of danger, perhaps, but were not really up to sustained flight.

Under those conditions, it might not seem much of a sacrifice to give up wings. There is always a tendency for land animals to take to sea life, since the sea is, on the whole, richer in life than land is. Besides this, the buoyancy of water makes life easier since one is not fighting gravity all the time; the temperature is more equable and is never either as hot or as cold as the land can get.

Thus, numerous land mammals have turned to the sea to a greater or lesser extent—whales, sea cows, seals, sea otters and so on. There are also sea turtles and sea snakes.

Among the birds, the penguins have turned their wings into paddles and can no longer fly, though they are accomplished swimmers.

It is rather surprising, perhaps, that 70 million years ago or so a bird had already done this, and even more thoroughly than the penguins have. The bird is *Hesperornis* (Greek for "western bird," because its fossils were found in the Americas). It, too, possessed teeth, but it had no keel on its breastbone. It had only the shriveled remnants of wings and propelled itself through the water with its large feet.

It was rather large for a bird, perhaps 5 feet long, but a sea creature is almost always larger than a land creature of the same type. The buoyancy of water means an organism doesn't have to pay for being large by requiring extra muscle to support the body against gravity. The added security that comes from being large and powerful is therefore particularly desirable. (That is why even the largest land animal that ever lived is only half the size of the largest sea animal that ever lived.)

Earlier than either Ichthyornis or Hesperornis is the skeleton of a bird, first discovered in 1861, of which there are only three samples known but which are perhaps the most important single fossils known to us.

It is the fossil of a creature about 3 feet long, with a head very much like that of a lizard (possessing teeth and no beak) and a long neck, again like that of a lizard, and a long tail, still like that of a lizard. There was no keel to its breastbone.

Wasn't it a lizard, then?

No, because it had feathers, imprints of which were clearly left in the rock. Those feathers are in a double row down the length of the tail, and are present all over the forelimbs. That is quite enough to make it a bird, and it is called *Archaeopteryx* (Greek for "ancient wing").

The English astronomer Fred Hoyle (b. 1915) recently claimed this fossil to be a hoax, with faked feathers—but paleontologists simply laughed at this. The details are so authentic, they could not have been faked, and all three fossil remnants of Archaeopteryx show the same thing.

Archaeopteryx lived in the later Jurassic and could be about 140 million years old. There seems to be some question as to whether it flew or merely glided, but most seem to think it could fly weakly.

Undoubtedly there were birdlike creatures before Archaeopteryx, and very recently a find was reported that might prove to be an example. Still, so far as we can tell now, the beginning of bird flight could not be very long before 140 million years ago and that may make bird flight twice as old as bat flight is.

Yet that was not the beginning of animal flight either.

As early as 200 million years ago, a group of reptiles developed flight without feathers. They were the *pterosaurs* (Greek for "wing-lizards"). The first pterosaur fossil was discovered in 1784. As in the case of bats, unwinged ancestors have not been found.

They had membranous wings like the bats, but whereas in bats the membrane stretched over all the fingers but the thumb, the pterosaurs had it attached to a vastly overgrown fourth finger. The first three fingers remained as small clawed digits outside the wing.

There seems to be some question as to how well the pterosaurs flew. No firm decision has been reached yet. Still, if the pterosaurs did fly, some paleontologists argue that they must have been warm-blooded and have had a hairlike covering as insulation. That matter, too, has not yet been resolved.

In any case, although some pterosaurs were no larger than sparrows, the largest ones were the largest flying animals there ever were. Toward the end of the Cretaceous, about 70 million years ago, the *Pteranodon* (Greek for "wing-no-teeth") flourished. It had a wingspan of up to 27 feet (8.25 meters), almost three times that of an albatross. To be sure, it was almost all wing. It may not have weighed more than 40 pounds (18 kilograms).

In 1971, however, remains of a pterosaur were located in Texas; its wingspan may have been as much as 50 feet

(15 meters), and it may hold the record for the weight of a flying animal.

At the end of the Cretaceous, 65 million years ago, all the pterosaurs died out quite suddenly, but birds survived.

And the pterosaurs, too, weren't the beginning of animal flight.

Mammals, birds, and reptiles may be widely different in some respects, but they resemble each other in having internal skeletons of bone. What's more, the skeletons resemble each other so much that it is quite clear that these three groups of animals are evolutionarily related and that all three descended from a common ancestor.

Mammals, birds, and reptiles may be lumped together (along with other organisms, such as fish) as *vertebrates*. This name is derived from a particularly important portion of the skeleton, the vertebral column or spine. The spine runs down the back of the animal and is a chain of individual, irregular bones called *vertebrae* (from the Latin, meaning "to turn," because the head turns on the upper vertebrae).

Vertebrates, together with a few more primitive creatures, make up a phylum (one of the grand divisions of the animal kingdom) called *Chordata,* because the most primitive internal skeleton consists of a rod called a *notochord* ("back cord"), and every chordate has had at least that at some time in its life.

Sometimes all the animals that are not vertebrates, including the most primitive chordates, are called *invertebrates,* but this is a biologically-useless term. The invertebrates are divided into about sixteen or so different phyla (there are always disagreements about the exact details of classification), and each one is as important from the standpoint of evolutionary mechanism as are the chordates.

One of the invertebrate phyla is *Arthropoda* (from Greek words meaning "jointed legs"). Arthropoda have external skeletons, or shells, and, as you might expect, jointed legs. Lobsters, crabs, and shrimp are examples of arthropods and, on land, spiders and centipedes. The biggest class of arthropods, however, is the insects, and they, in fact, are the most numerous, the most cleverly adapted, the most various, and the most successful of all forms of life.

There are more species of insects alive than of all other forms of life put together. And of the millions of species of life that may exist undiscovered in out-of-the-way places of the world, the vast majority are probably more insects.

There may be 2 million species of insects all together, as compared with about 4,000 species of mammals. Insects are, by and large, very short-lived and can give birth to incredible numbers of young. This means that for them evolution can proceed at a breakneck speed, and many species will evolve.

Insects first made their appearance in the fossil record long before the Mesozoic, perhaps as long as 350 million years ago, and by then they already had wings. There are some very primitive wingless insects that survive even today, and they probably extend the evolutionary history of insects even farther back.

Whereas the wings of reptiles, birds, and mammals, however different in detail, are all modifications of the forelegs, the insects have wings that have no relation to their legs at all. The wings are, instead, thin, stiffened extrusions of the material making up their skeletons.

Insect wings are far flimsier in structure than the wings of vertebrates, and insects pay the price for it by being unusually small. There are indeed some comparatively large insects. The Goliath beetle can be nearly 6 inches (15 centimeters) long and weigh nearly 4 ounces (about 90 grams), so that it is considerably larger than the smallest mammals and birds, but this is most exceptional. By far the largest majority of insects are small (think of house-flies) or even minute (think of gnats). The smallest insects are barely large enough to see with the unaided eye.

The insects were the first animals capable of true flight, which means that true flights began about 350 million years ago, and for two-fifths of that time the only fliers were insects.

However, putting the insects to one side, let's return to

birds and mammals. Both birds and mammals are clearly related to the reptiles and the more primitive the bird or the mammal, the more reptilian are its characteristics. From this it is easy to deduce that both birds and mammals evolved from reptiles.

Earlier than 150 million years ago, there are no birds or mammals in existence, but reptiles were flourishing. Let us, therefore, turn to them and take up the matter of their beginnings.

# 10

## REPTILES

During the Mesozoic—the high noon of the reptiles—there flourished several important sub-classes of this group, sub-classes most easily distinguished by differences in the skull structure. Actually, paleontologists had little choice in choosing ways of differentiating among the reptiles. It is almost always the bones that have survived in fossil form and, in particular, the skulls.

Nor are the differences in skull forms to be dismissed as trivial. Slight changes in the structure of the skull are usually accompanied by other changes in the skeleton that are indicative of important differences in appearance and way of life. We find this to be so among the different kinds of reptiles that have survived today, and there is no reason to think matters were different in the past.

Thus, the reptiles are divided into sub-classes depending on the number and position of holes on either side of the skull just behind the eye socket, holes that allow room for jaw muscles to pass through and to swell on contraction.

There are reptiles with no such holes at all, and they belong to the sub-class of *Anapsida* (from Greek words meaning "no openings"). Such reptiles can be referred to, familiarly, as anapsids.

There are three sub-classes of reptiles with a single hole behind the eye socket on each side. Paleontologists distinguish among the three, according to the position and size of the hole, and the precise arrangement of the bones about it. These three sub-classes are *Synapsida* ("with opening"), *Parapsida* ("side opening"), and *Euryapsida* ("wide opening").

Finally, there is a sub-class with two holes behind each eye socket, and this is *Diapsida* ("two openings"). The diapsids are divided into two groups, based on differences in the teeth, the sub-groups being *Lepidosauria* ("scaly lizards") and *Archosauria* ("ruling lizards").

The archosaurians were the most successful of all the groups of Mesozoic reptiles and were subdivided into five orders. One of the orders is *Saurischia* ("lizard hips"). This name is given them because in all the members of this order the bones of the hips are arranged rather like the bones in the hips of modern lizards. A second order is the *Ornithischia* ("bird hips") because here the hipbones are arranged as in modern birds.

The saurischians and the ornithischians, taken together, are the animals popularly called dinosaurs. The word *dinosaur* ("terrible lizard") was first coined by the English zoologist Richard Owen (1804–1892) in 1842. At that time, little was known about these reptiles, and it was not clearly understood that they existed in two groups that were distinctly different from each other. The word *dinosaur* is therefore not an official zoological classification nowadays, but the word can never be wiped out of popular usage. Even scientists use it as a brief way of referring to these two groups.

The saurischian dinosaurs had their heyday first. They are divided into two suborders, *Theropoda* ("beast feet") and *Sauropoda* ("lizard feet") because the toebones of the former more closely resemble those of mammals in number, while the toebones of the latter more closely resemble those of lizards. In addition, the theropods are bipeds, tending to walk on their hind legs only, while the sauropods are quadrupeds and walk on all fours.

Many of the early theropods were quite small. One of them, *Compsognathus* ("elegant jaw," because the skullbones were so small and delicate), lived about 150 million years ago and was no larger than a chicken. It is the smallest known dinosaur. Toward the end of the Mesozoic, there were bipeds of this sort that looked almost exactly like ostriches except that they had scales instead of feathers, and small forelimbs with clutching paws instead of useless wings.

Some of the theropods grew enormous, however, and became the *carnosaurs* ("meat lizards," because they were meat eaters). The best-known of these is *Tyrannosaurus Rex* ("master lizard, the king"), which, along with other carnosaurs that may have been even larger, were the most fearsome land-carnivores that ever existed.

The total length of a large carnosaur may have been up to 50 feet (15 meters), and the total weight about 7 tons. This would be over eight times the weight of a modern Kodiak bear, the largest land carnivore now alive. The head of a large carnosaur was 4 feet (1.2 meters) long with teeth 6 inches (15 centimeters) long, and it towered 16 feet (nearly 5 meters) above the ground. The carnosaurs were bipedal, too, and the forelimbs were small compared to the rest of the body, so that they looked like giant kangaroos. The enormous thighs of these reptiles showed that they just about reached the limits of size for a land animal supported on two legs.

The sauropods may also have been of bipedal ancestry. Although they walked on all four legs, the forelimbs were usually shorter than the hindlimbs so that the back of the sauropod generally sloped upward from shoulders to hips.

To the average person, these sauropods are the most familiar of the dinosaurs, and the very word *dinosaur* calls up their image. They were super elephantine in structure, with long necks at one end and long tails at the other. Indeed, they looked like enormous snakes that had swal-

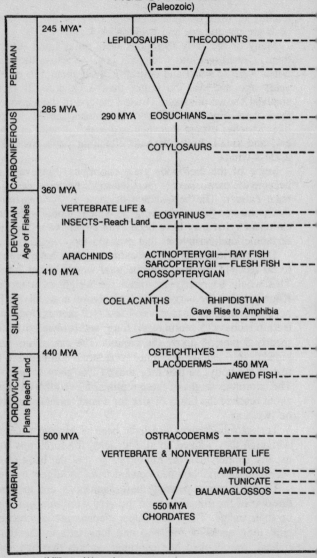

# AGE OF AMPHIBIA
### (Paleozoic)

| | | |
|---|---|---|
| PERMIAN | 245 MYA* | LEPIDOSAURS    THECODONTS - - - - |
| CARBONIFEROUS | 285 MYA | 290 MYA   EOSUCHIANS - - - - |
| | | COTYLOSAURS - - - - |
| DEVONIAN Age of Fishes | 360 MYA | VERTEBRATE LIFE & INSECTS-Reach Land   EOGYRINUS - - - - |
| | | ARACHNIDS |
| | | ACTINOPTERYGII—RAY FISH |
| | | SARCOPTERYGII—FLESH FISH - - - |
| | 410 MYA | CROSSOPTERYGIAN |
| SILURIAN | | COELACANTHS    RHIPIDISTIAN |
| | | Gave Rise to Amphibia - - - - |
| ORDOVICIAN Plants Reach Land | 440 MYA | OSTEICHTHYES - - - - |
| | | PLACODERMS—450 MYA |
| | | JAWED FISH - - - - |
| CAMBRIAN | 500 MYA | OSTRACODERMS - - - - |
| | | VERTEBRATE & NONVERTEBRATE LIFE |
| | | AMPHIOXUS - - - - |
| | | TUNICATE - - - - |
| | | BALANAGLOSSOS - - - - |
| | | 550 MYA CHORDATES |

*Millions of Years Ago

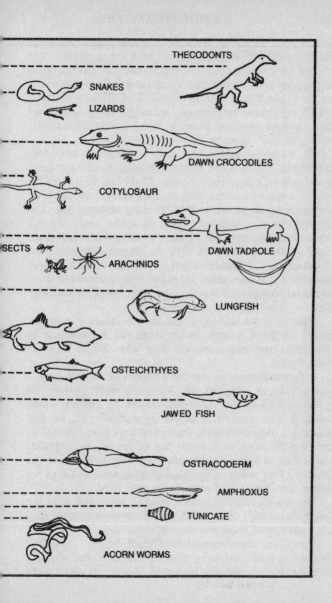

THECODONTS

SNAKES

LIZARDS

DAWN CROCODILES

COTYLOSAUR

DAWN TADPOLE

SECTS

ARACHNIDS

LUNGFISH

OSTEICHTHYES

JAWED FISH

OSTRACODERM

AMPHIOXUS

TUNICATE

ACORN WORMS

lowed giant elephants, with the columnar legs of the latter breaking through and walking off with the creature.

The large sauropods were vegetarians. In general, plant eaters can grow larger than meat eaters, for the world is richer in plant food than in animal food. Thus, elephants, which are strictly herbivorous, are larger than grizzly bears that eat meat also, and these, in turn are larger than tigers, which eat only meat.

The longest of all the sauropods was *Diplodocus* ("double beam," from some skeletal details). Some specimens seem to have been nearly 90 feet (27 meters) long from the snout to the end of the long tapering neck, past the body proper, and on to the end of the long tapering tail. The Diplodocus was slenderly built, however, and may not have weighed more than 11 tons, which would make it not very much more massive than the largest elephants. The *Brontosaurus* ("thunder-lizard," perhaps because it was imagined that the noise it made was thunderous as it clumped along) was shorter, but more massive, and may have weighed up to 35 tons.

More massive still was the *Brachiosaurus* ("arm lizard," so-called because in the course of its evolution its forelimbs had lengthened till they were longer than the hind limbs).

A Brachiosaurus was about 75 feet (23 meters) long—not as long as a Diplodocus, but much more massive. Its head towered 40 feet (12 meters) above the ground, which made it well over twice as tall as a giraffe, or, for that matter, a Baluchitherium. Its weight may have reached as much as 80 tons, eight times that of the largest elephant, twice that of a Baluchitherium—but only half that of the largest living whale. The Brachiosaurus was, as far as we know, the largest land animal that ever lived.

The ornithischian dinosaurs reached their peak after the saurischians did, and toward the end of the Mesozoic Era, they developed some spectacular armored types.

There was the *Stegosaurus* ("roof lizard," so-called because it possessed bony plates that at first were thought to plate its back like tiles on a roof). Later, the plates were thought to line the back in a double-row, each standing on end. Very recently, evidence has been presented to the effect that they were present in a single row.

The Stegosaurus showed clear signs of ancestral bipedality, for its front legs were little more than half the length of the hind legs. It is usually considered a particularly brainless animal, for its tiny head contained a brain no larger than a modern kitten's, even though its body was 30 feet (9 meters) long and more massive than an elephant. It became extinct in the early Cretaceous, about 120 million years ago, probably before the giant carnosaurs appeared on the scene. The famous sequence in Walt Disney's production *Fantasia* in which a Tyrannosaurus attacks and kills a Stegosaurus, is very likely anachronistic.

*Ankylosaurus* ("crooked lizard") evolved later than the Stegosaurus and was indeed contemporary with the carnosaurs. It was probably the most heavily-armored creature of all time. It was about the size of a Stegosaurus but was lower and broader so that it could not be easily overturned to expose its unarmored belly. Its back, from skull to tail, was layered with massive bony plates that were drawn into strong spikes along their sides. The tail ended in a bony knob that probably had the force of a battering ram when swung. It was virtually a living tank, and perhaps even a carnosaur would have thought twice about challenging it.

Then there is *Triceratops* ("three horned"), which was built like a super-rhinoceros. It was smaller than Stegosaurus and Ankylosaurus, and its armor was concentrated in its head region. A broad frill of bone, 6 feet (1.8 meters) across, extended backward from the head and covered the neck. The face bore three horns, two long sharp ones above the eyes, and a shorter, blunter one on the nose. In addition, the mouth was equipped with a strong, parrotlike beak.

Then, at the end of the Cretaceous, 65 million years ago, something happened: All the saurischians and ornithischians that were then alive—all the reptilian dinosaurs without exception—died in what seems to have been a short period of time, geologically speaking.

However, the dinosaurs made up only two of the orders of the archosaurian subclass. There were three other orders.

Of those three, one gave rise to the pterosaurs, which we mentioned in the previous chapter. The pterosaurs, though they lived in the time of the dinosaurs, and although grouped with them under Archosauria, were *not* dinosaurs because they were not members of the only two orders that are granted that unzoological name.

Nevertheless, at the end of the Cretaceous, when the dinosaurs died out, the pterosaurs died out also.

A fourth order of the Archosauria is *Crocodilia*. Just before the end of the Cretaceous, there was *Deinosuchus* (''terrible crocodile''), which was the largest crocodilian we know of. It was 50 feet (15 meters) long. It didn't survive the Cretaceous, but some smaller members of the order did. Crocodiles and their relatives, the alligators and the caymans, are still alive today.

Of the orders of reptiles alive today, the Crocodilia are the only members of Archosauria and, although not dinosaurs, are the closest reptilian relatives to the dinosaurs still alive.

The last order of the Archosauria was, in some ways, the most remarkable, for it gave rise to Archaeopteryx, and through it to the birds. The birds, like Crocodilia, survived the Cretaceous and are also examples of Archosauria. They are as closely related to the dinosaurs as the crocodiles are, but birds have evolved away from reptilian characteristics in so dramatic a fashion (as, for example, in feathers, flight, and warm-bloodedness) that they are not considered reptiles at all.

As I mentioned earlier, there is another subclass of Diapsida in addition to Archosauria. This is the Lepidosauria. During the Mesozoic, the lepidosaurs lagged far behind the archosaurs in importance. Two orders of lepidosaurs, however, survived the mass extinction at the end of the Cretaceous. One is *Squamata* (''scaly''), from which have descended the snakes and lizards of today—the most successful of the living reptiles.

The largest living lizard is the *Komodo dragon,* which is

found on Komodo and a few neighboring islands in Indonesia. A large Komodo dragon can be up to 10 feet (3 meters) in length and weigh up to 365 pounds (165 kilograms). To a startled onlooker seeing it for the first time, it might appear to be a small dinosaur—but, of course, it isn't.

Another order of the Lepidosauria is *Rhynchocephalia* ("snout heads," because they have prominent, beaky snouts). This order was never important, and it survived the mass extinction of the reptiles by the narrowest possible margin. A single, rare species survives.

This survivor is a moderately large lizardlike creature, about 2.5 feet (0.75 meters) in length. It is now found only on a few offshore New Zealand islets, where it is sternly protected by law. Its common name is *tuartara* ("back spine" in Maori, since in addition to the scales that cover its body, it has a line of spines down its backbone). Its more formal name is *Sphenodon* ("wedge tooth"). Though it looks like a lizard, it differs from lizards in a number of ways. For one thing, it has a particularly well-developed pineal gland at the top of its brain, a gland not nearly as well developed in lizards, or in other vertebrates either. In the young sphenodon, it bears the anatomical appearance of a third eye, though there is no indication that it is light-sensitive.

Now let's pass on to the three reptilian orders with only one opening in the skull on each side behind the eye socket. Of these, the Euryapsida included large marine reptiles that flourished in the Mesozoic. These are known as *plesiosaurs* ("near-lizards"), and they are very much like dinosaurs in outer appearance. Some of them look like sauropods, with four long flippers in place of four long legs. One of them, the *elasmosaurus* ("plated lizard"), had a neck about 20 feet (6 meters) long with seventy vertebrae along its length, compared to our seven. It was the longest neck ever found on any animal. (Some people think that the so-called Loch Ness monster is a miraculously surviving plesiosaur, but I think the chances are just about zero that the Loch Ness monster exists at all.)

The Parapsida gave rise to marine reptiles, too, and in their case the adaptation was more extreme. The parapsids that are best known are the *ichthyosaurs* ("fish lizards"), who looked very much like reptilian dolphins. They gave birth to living young, but without placentae, as sea snakes do today. One way in which they differed from dolphins is that the fluked tail of the ichthyosaur was vertical, whereas in dolphins it is horizontal. The ichthyosaur spine ran down into the lower lobe of the tail, rather than remaining centered as in the dolphin. Some ichthyosaurs were as much as 25 feet (7.5 meters) long, but their brains were much smaller than those of dolphins.

Both plesiosaurs and ichthyosaurs are now extinct. The plesiosaurs died out at the end of the Cretaceous along with the dinosaurs, but the ichthyosaurs seem to have died out 90 million years ago, well before the end of the Cretaceous.

The remaining "one-hole" order are the Synapsida. They are among the earlier reptiles and evolved even before the Mesozoic. They would not be considered very remarkable or noteworthy except for one thing. They developed mammallike traits. One of the suborders, the *theriodonts* ("beast-teeth"), developed a skeleton that was quite mammalian in many respects. For instance, their teeth were far more mammalian in character than reptilian (as the name of the suborder implies). At some point, the theriodonts may even have developed warm-bloodedness and hair, though there is no way of telling this from the fossil remains.

To those of us who assume that mammals are "superior" to reptiles, it might seem that the Synapsida would be very successful. It might seem that every further development of a mammalian trait would give the synapsids that much more of an advantage over the other reptilian orders.

That, however, does not seem to have been so. All the synapsids died out early. Even the mammallike theriodonts were mostly gone by 170 million years ago, less than halfway through the Mesozoic, leaving the dinosaurs triumphant. However, some small theriodonts survived, having become particularly mammalian. Because of the paucity of fossil remains and the gradual nature of the change, it isn't possible to say that at exactly one particular point a creature developed that was a true mammal. In any case, it was not the mammalian characteristics that ensured survival, but the fact that the first mammals were so small. Between escaping notice and being able to scurry quickly into shelter, they avoided destruction by the reptiles—until such time as the reptiles themselves died, a hundred million years later, and gave the little mammals their chance.

One last reptilian order, the *Anapsida,* with no holes at all behind the eye-sockets, are in some ways the most primitive of the reptiles, and they, too, got their start well before the Mesozoic. Oddly enough, they managed to survive the end of the Cretaceous while more advanced reptiles did not. The turtles and tortoises of today are living examples of the Anapsida.

But why was there so much extinction at the end of the Cretaceous? Why did so many of the large reptiles die then, after 150 million years of successful development?

Many solutions have been offered. It was suggested that perhaps new forms of plant life evolved and flourished, plants the herbivorous dinosaurs couldn't chew or digest. When they died, the carnivorous dinosaurs who lived on them died also.

Or else there might have been climatic changes. Perhaps a period of glaciation cooled the ocean drastically, perhaps a change in the land-sea configuration resulted in the disappearance of coastlines, or perhaps a fall in sea level drained the shallow seas. Perhaps there was the coming of a new disease, or a nearby supernova drenched the Earth in a bath of cosmic rays. It was even suggested that the small mammals learned to live on dinosaur eggs.

Then, in 1979 an American scientist, Walter Alvarez, was analyzing long cores of sedimentary rock, obtained in Italy, by a very delicate chemical technique called "neutron activation analysis." He was hoping to work out something about the rate at which the sedimentary rock was laid down over long periods of time. That didn't work, but, to his surprise, Alvarez and his coworkers discovered that there was a thin section of the sedimentary rock in which the rare metal, iridium, was twenty-five times as high as it was either above or below. The iridium had appeared in unusual quantity (still very little, to be sure) at one particular time, and this time turned out to be exactly at the end of the Cretaceous.

There had to be a connection. Iridium is a very rare metal everywhere in the Universe as far as we know, but it is particularly rare in the Earth's crust, because what iridium is present on Earth is mostly in the Earth's core of molten iron. Meteors, for instance, are known to be richer in iridium than the Earth's crust is (though not than the Earth as a whole is).

Further investigation showed that the iridium layer is widespread over the Earth. The thought arose, then, that 65 million years ago an asteroid or, more likely, a comet, perhaps several miles across, must have struck Earth and created enormous earthquakes, volcanic eruptions, and tidal waves. In addition, it would have kicked enough dust into the upper atmosphere to block virtually all sunlight for an extended period of time, thus killing plant life and cutting off the food supply for animal life.

Most of life on Earth would die in consequence. The larger animals would be particularly vulnerable, since there were fewer of them and they required more food per individual. Small animals had a better chance to endure since they could live on the corpses of the large animals who died or, if herbivorous, on seeds, stems, bark, and other surviving remnants of plants. While large animals would be wiped out, small animals would survive or not survive at least partly as a matter of random chance.

In any case, once the Earth settled down, those plants and animals that survived would find themselves on a relatively empty Earth and could evolve quickly into a multiplicity of species again.

The Cretaceous mass extinction is the most famous example of the sort because it brought an end to the dinosaurs, a group of animals who have a strong grip on the imagination of people. That was not the only mass extinction, however. In fact, some paleontologists, studying the fossil record carefully, maintain that such mass extinctions come about every 26 million years.

Naturally, they are not always extreme. Sometimes they are relatively mild, but one, at least, which marked the end of the era preceding the Mesozoic, was even worse than that which ended the Mesozoic. Some 95 percent of all existing species were wiped out during the Permian extinction.

Are all extinctions brought about through bombardment of Earth from outer space? If so, why should those bombardments come every 26 million years? One suggestion is that the Sun has a small companion star, far out in space, that revolves around it every 26 million years. At one end of its orbit, it is so far away it affects nothing at all, but at the other end, which it reaches every 26 million years, it comes close enough to the Sun to pass through a cloud of 100 billion small icy comets that are thought to lie beyond Pluto's orbit. These are disturbed, and a few million comets may plunge toward the inner Solar system, some inevitably striking the Earth.

If this is so, Earth is continually getting a new start in evolution. This resembles Bonnet's *catastrophism*, which was mentioned earlier in the book, but only distantly. The new catastrophism describes these times of dread as being separated by far longer periods of time than Bonnet imagined, and so far at least, none of them have wiped out life altogether, as Bonnet's were supposed to. In the new catastrophism, each new step of development arises through the further evolution of the survivors of the catastrophe. In Bonnet's system, each new step required divine creation from scratch.

The explanation of mass extinctions by bombardment from outer space is still highly controversial, and many paleontologists simply will not accept it. They do not feel that the mass extinctions are truly periodic, and they tend to advance other reasons, such as the cooling of the Earth during an Ice Age, for the extinctions.

Even if the notion of periodic extinctions through cometary bombardment turns out to be correct, the next scheduled mass extinction is about 15 million years from now, and there's no reason for immediate anxiety about it.

Now we can return to the matter of the beginning of the reptiles. I have already said that the Synapsida and the Anapsida evolved before the beginning of the Mesozoic. Let us therefore consider the period of time that preceded the Mesozoic: the earliest of the three major periods in which fossil remnants are prominent. This earliest period of fossilization is the Paleozoic (''ancient animals''). The Paleozoic lasted 355 million years altogether, so that it is longer than the Cenozoic and Mesozoic Eras put together.

The Paleozoic is divided into six periods, which, going from the most recent to the most ancient, are as follows:

*Permian* (from a province in eastern Russia once known as Perm, where rock layers from this period were first studied). It was at the end of this period that the worst mass extinction ever brought the Paleozoic to an end and allowed the relatively few survivors to evolve into the life of the Mesozoic.

*Carboniferous* (''coal bearing,'' because much of the coal we mine appears in rocks from this period).

*Devonian* (from Devonshire in southwestern England, where these rocks were first studied).

*Silurian* (named for a tribe in southern Wales in Roman times, since these rocks were first studied in southern Wales).

*Ordovician* (named for another Welsh tribe).

*Cambrian* (named for Wales itself, which was known as Cambria in Roman times).

For the moment, let's pin down only the time spans of the first two periods. The Permian endured from 245 MYA back to 285 MYA, a duration of 40 million years. The time between the Permian extinction 245 million years ago and the Cretaceous extinction 65 million years ago is 180 million years, which is equal to seven times the 26-million-year period that has been suggested as separating one mass extinction from another.

The *Carboniferous* extended from 285 MYA back to 360 MYA, a duration of 75 million years.

The early reptiles that were in existence in the Permian suffered greatly in the Permian extinction, and many of their species died, especially among the Synapsida or mammallike reptiles (though obviously some survived).

Appearing shortly after the Permian extinction, about 240 million years ago, were the *thecodonts* ("socket teeth"). Having teeth rooted in sockets is characteristic of the archosaurs, so the thecodonts were, in fact, the first archosaurs.

Some of the thecodonts had legs sprawled out to the side, as in modern lizards, which made for clumsy walking. Others, however, had legs under the body as was true of the dinosaurs. Some of the thecodonts were lightly built and had long hind legs, indicating that they might be able to run bipedally, and these were almost dinosaurs. Another thecodont that lived about 200 million years ago seems to have had enlarged, loosely overlapping scales that might represent the first movement in the direction of feathers.

The thecodonts survived into the early Jurassic, about 193 million years ago, when another mass extinction carried them off. However, by that time they had left descendant species that survived, and it was from these that the dinosaurs, pterosaurs, crocodilians, and birds arose.

If the thecodonts were the first archosaurs, they were certainly not the first reptiles. They descended from certain reptiles that had survived the Permian extinction. These are the *eosuchians* ("dawn crocodiles"), which first arose about 290 million years ago in the late Carboniferous. They did remarkably well, some of them surviving even the Cretaceous extinction and not dying out altogether till about 50 million years ago in the Eocene, at the time of another period of mass extinction.

Early on, some of the eosuchians evolved into thecodonts and others into the lepidosaurs who were the ancestors of the tuartara, the lizards, and the snakes. The eosuchians were the first reptiles with a diapsid skull, though their teeth remained primitive.

The eosuchians were descended from the *cotylosaurs* ("cup lizards," so-called because their vertebrae are cup-shaped). The cotylosaurs may have come into existence 300 million years ago, in the late Carboniferous. It is they who seem to be the original reptiles from which all other reptiles (and birds and mammals as well) are descended. The cotylosaur skull is anapsid as are those of turtles and tortoises today.

What is most important about the cotylosaurs, and what most characteristically divides the reptiles generally from vertebrates that evolved earlier, are the eggs that reptiles lay. Animals more primitive in this respect than reptiles must lay their eggs in water, since those eggs would quickly dry if laid on land and would then die. That meant that the ancestors of reptiles had to lead at least the early parts of their lives in water.

The cotylosaurs had evolved a protected egg, capable of being laid on land. In the first place, the egg is surrounded by a protective shell of thin limestone (calcium carbonate), which is permeable to air but not to water. Air can reach the developing embryo inside, but water cannot leave it. The embryo develops in a small pool of water preserved inside the egg, with an elaborate series of adaptations to allow the growing embryo to get rid of wastes, which are tucked into other membranes.

The reptilian egg, as developed by some primitive cotylosaurs about 300 million years ago, made all subsequent land life of vertebrates (including reptiles, birds, and mammals) possible. The reptilian egg was, therefore, the most important vertebrate "invention" in reproduction, not to be matched until the "invention" of the placenta by the more advanced mammals about 230 million years later.

But though total land life is possible for reptiles and their descendants, it is clear that reptiles must have been descended from more primitive animals that lived in water at least part of the time. We ought to ask, then, what was the beginning of land life of any kind?

# 11

## LAND LIFE

Life in the water is, in some ways, very easy. Water is buoyant and supports living things, at least to a considerable extent. Sea life does not have to fight gravity; it lives in a three-dimensional world, able to move easily not only forward and backward, and left and right, but also up and down.

To be sure, flying animals live also in a three-dimensional world, but flying through air takes much more energy than swimming through water. In order to fly, birds and bees (and, just possibly, pterosaurs, too) must be warm-blooded, maintaining a high metabolic rate—that is, producing energy at a high level. Insects, which are cold-blooded, compensate by being so small that the lesser buoyancy of air is sufficient to relieve them of at least some need to support their weight.

In the sea, on the other hand, living organisms may be cold-blooded and, at the same time, large. They may swim slowly and, so to speak, lazily, without falling, whereas birds must maintain a good speed at the cost of considerable energy expenditure if they are to remain airborne.

Even large birds who, by making use of air-currents, can soar for long periods with scarcely any energy expenditure, must expend a large amount of energy to gain altitude initially.

Then, too, in the sea, temperatures do not vary widely, and for the most part the environment is stable. What's more, water is absolutely essential to life and the ocean is 96.7 percent water.

The ocean is, in fact, so benign an environment that this very quality can be a terrible disadvantage under certain circumstances. Organisms that live in the warm tropical oceans are adapted to the unvarying kindness of the sea. But then, when the temperature of the tropical oceans drops somewhat, for instance, as a result of an Ice Age, the life forms find they cannot tolerate the change. Tropical marine life seems to suffer to an unusual extent in episodes of mass extinction, presumably because of its vulnerability to cooling.

Still times of mass extinctions make up a tiny percentage of the total stretch of time during which organisms have lived on Earth. For many millions of years at a time, the ocean environment has remained stable and life has continued essentially unruffled.

It would seem, then, that there is little to lure living things from water to the dry land.

To emerge from water and live on the surface of the dry land, living things would have to evolve mechanisms to prevent desiccation, and they would have to be able to endure temperatures that are liable to be, at times, much higher, or much lower, than they would encounter in the sea. They must be able to endure environmental factors such as direct sunlight, rain, snow, and wind. To make progress, they must either wiggle or crawl slowly over a two-dimensional surface or else develop limbs that will be strong enough to lift them clear of the ground under the pull of a gravity undiluted by the buoyancy of water.

Nor is this all. In the sea, there is oxygen that is dissolved in the water. This oxygen can be absorbed by a sea organism through organs called *gills* that are richly supplied with blood vessels. The water passes ceaselessly over the gills, and oxygen diffuses from the seawater into the blood. In the sea, also, waste products (which may, in themselves, be poisonous) can be excreted into the water as soon as they are formed, and there they are diluted to harmlessness as they undergo chemical and biological changes that prevent them from ever accumulating in dangerous amounts.

On the land, however, oxygen must be obtained from the air and must be dissolved in the moisture lining the interior of the lung before it can be used—and that moisture must be maintained and never allowed to dry out. This is a much more complicated system than is required in water.

Then, too, wastes in land animals cannot be voided steadily, since that can only be done by having them in water-solution and that would waste too much precious water. The land animal would be dried and dead almost at once. Instead, wastes in land animals must be allowed to accumulate to some extent, must be converted into products that are not too toxic, and must then finally be eliminated with a minimum of water.

In addition, the sea is full of life, which means full of food, while the dry land (even today, and much more so hundreds of millions of years ago) is comparatively barren.

Why, then, should life in the sea evolve all sorts of very complicated adaptations that would fit it for life on land, when life in the sea is so much easier and better?

Evolution, you must understand, is not a matter of purposeful change. Life didn't "want" to move onto dry land.

The fact that life in the sea is so easy means that it swarms with life forms that eat and are eaten. The competition is fierce. At the tidal rims of the ocean, living organisms will, usually, avoid penetrating too far up the slope of the shore since, the higher they move as the tide advances, the greater the chance of being accidentally exposed to the killing absence of water, as the tide recedes.

If a particular organism, however, happens to be able to

survive a short period without water-cover, it can live higher up the slope than other organisms do and be somewhat more secure from predation while there. It will also have less competition in finding such food as does exist there. One can then imagine a kind of leap-frog series of adaptations, where organisms survive better if they can endure the absence of water for longer and longer periods, with one organism gaining an advantage, and then another. This doesn't happen rapidly, of course, but over a period of millions of years you can end by having organisms that are well adapted to life on land for at least substantial periods, if not permanently.

Then, again, organisms that live in constricted bodies of water may find that at times the water grows brackish, and the dissolved oxygen runs low. In such an emergency, any organisms that can gulp air and extract oxygen from it can survive such a brackish period, and that gives them an advantage. Some fish have developed primitive lungs for the purpose.

Also, organisms that live in pools may find that during a drought, a pool may dry up to the point where there is no room for its load of life. Any organism that can manage to wiggle or crawl from that pool into a nearby larger one can better survive. If its fins or paddles are strong enough to support it during this journey, even if only clumsily, so much the better.

Until the Paleozoic was about two-thirds over, all of life existed in water and the land was barren. The most advanced vertebrates then alive were the fish (which still dominate the oceans today).

However, the stresses that arose in the sea resulted in the evolution of fish that could withstand sunlight, that could avoid desiccation, that possessed lungs and legs, and so on.

For a long time after the rise of these land-living vertebrates, one kind of land adaptation was missing. The vertebrate egg (prior to the development of the reptilian egg by the cotylosaurs) could not survive on land. No

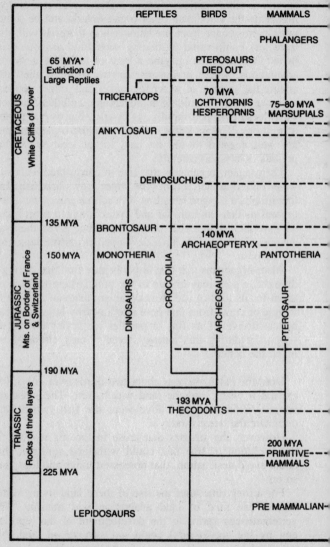

# AGE OF REPTILES
(Mesozoic)

| | | REPTILES | BIRDS | MAMMALS |
|---|---|---|---|---|

PHALANGERS

65 MYA*
Extinction of
Large Reptiles

PTEROSAURS
DIED OUT

TRICERATOPS

70 MYA
ICHTHYORNIS
HESPERORNIS

75–80 MYA
MARSUPIALS

ANKYLOSAUR

CRETACEOUS
White Cliffs of Dover

DEINOSUCHUS

135 MYA

BRONTOSAUR

150 MYA

MONOTHERIA

140 MYA
ARCHAEOPTERYX

PANTOTHERIA

JURASSIC
Mts. on Border of France
& Switzerland

DINOSAURS

CROCODILIA

ARCHEOSAUR

PTEROSAUR

190 MYA

193 MYA
THECODONTS

TRIASSIC
Rocks of three layers

200 MYA
PRIMITIVE
MAMMALS

225 MYA

LEPIDOSAURS

PRE MAMMALIAN

*Millions of Years Ago

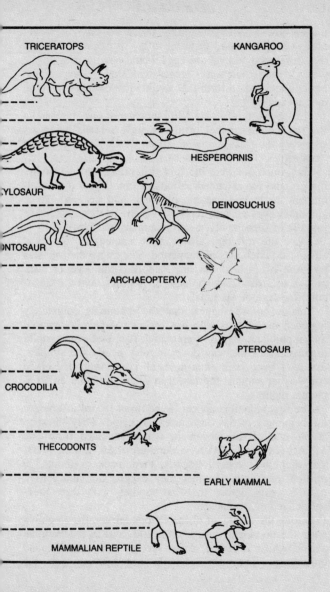

TRICERATOPS

KANGAROO

HESPERORNIS

ÇYLOSAUR

DEINOSUCHUS

ONTOSAUR

ARCHAEOPTERYX

PTEROSAUR

CROCODILIA

THECODONTS

EARLY MAMMAL

MAMMALIAN REPTILE

matter how a vertebrate might thrive on land, it would always have to return to water to lay its eggs. The young that developed from those eggs would have to stay in the water during their early stages and slowly develop the legs, lungs, and so forth that would enable them to live on land as adults.

This necessity of living a life in water at one stage and a life on land in another results in such animals being assigned to the class *Amphibia* (from Greek words meaning "both lives").

The Amphibia were the first vertebrates capable of living on land for extended periods of time. With the development of an egg capable of being laid on land, some amphibia evolved into reptiles, which, in turn and in the fullness of time, evolved into mammals and birds.

Thus, the different eras can be named for the most advanced vertebrates of the period. The mid-Paleozoic was the Age of Fishes, the late-Paleozoic the Age of Amphibia, the Mesozoic, the Age of Reptiles, and the Cenozoic, the Age of Mammals.

This is not to suggest that the mammals completely replaced the reptiles who had earlier completely replaced the amphibians. Reptiles, amphibia, fish, and even simpler organisms, all the way down to very nearly the simplest that ever lived, still exist now, all competing, and all in one way or another successful in some particular environmental niche.

The first amphibia appear in the fossil record just before the beginning of the Carboniferous period. They show up at the end of the Devonian, which extended from 360 MYA back to 410 MYA, a duration of 50 million years. The amphibian record extends, then, back to about 370 million years ago, so that they existed on land for 70 million years before the first reptiles with their land-specialized eggs appeared.

During the early part of the Carboniferous, the amphibia were the dominant form of land life, and in the Permian period which followed, some species were armored and quite large. They did not look very different from the primitive reptiles that were soon to develop. The largest known amphibian was *Eogyrinus* (Greek for "dawn tadpole," though it looked far more like an alligator than a tadpole). It grew up to 15 feet (4.5 meters) long.

As the reptiles evolved, however, the large amphibia declined and were extinct by the end of the Triassic. By that time amphibia of the modern type were evolving; these found survival value not in size and armor, but as the early mammals did, in smallness and obscurity. Modern amphibia are small animals generally—frogs, toads, salamanders, and legless caecilians. The largest amphibian species now alive is the Chinese giant salamander, which is 3 feet (1 meter) long, though there are reports of individuals as much as 5 feet (1.5 meters) long.

Vertebrate land life began, then, 370 million years ago with the earliest amphibia, but there was life on land to greet them because the arthropods succeeded in colonizing the dry land before the vertebrates did. The arthropods had a number of advantages that made it possible for them to do so.

For one thing, arthropods are small, in general, and those species that emerged on land were particularly small so that gravity was not much of a factor.

For another, arthropods, unlike vertebrates, have an external skeleton made of chitin, a substance quite unlike the vertebrate bone. Chitin, in fact, is chemically more closely related to the cellulose that is the characteristic constituent of wood. Whereas cellulose is built up of sugar units, however, chitin has those sugar units plus nitrogen-containing groups as well. Chitin is horny, tough, and rather flexible. It serves to protect arthropods under water, and the protection endures on land, serving to mitigate the effects of sunlight and to slow the drying process.

Furthermore, bottom-dwelling arthropods had developed chitin-clad limbs that were stiff enough and strong enough to lift them clear of the sea bottom, with the aid of the buoyancy of water. And since they were small, those same limbs would support them on land against the pull of gravity.

Again, for the smaller arthropods the problems of obtaining oxygen and of disposing of waste were easier to solve.

The insects were, of course, the most successful arthropods, but we have little in the way of fossil information in the case of these small and fragile organisms. The largest known insect that ever lived was a dragonfly that flourished before the end of the Cretaceous and had a wing span of up to 2.25 feet (0.66 meters). However, it was almost all wing, and the body itself was not at all massive.

Primitive wingless insects (and some, like the "springtails," exist even today) may have reached land 370 million years ago, at about the time the vertebrates were emerging, but that was the second arthropod invasion.

The first arthropod invasion included the arachnids (spiders and scorpions, for instance, which differ from insects most noticeably in having eight legs rather than six, two segments rather than three, and no wings). To these, we may add some non-arthropods, such as snails and earthworms. The first primitive animals of this type to venture onto land may have taken the step some 400 million years ago, at the very beginning of the Devonian.

The first amphibia that invaded the land found themselves, then, in an environment within which various small creatures that had been multiplying and diversifying for 30 million years were flourishing. We can therefore picture these amphibia feeding on insects and so on. (In fact, modern frogs still live on insects.)

But what would the insects and other small organisms feed on? Each other?

That is not a way of solving the food problem in the long run, for feeding does not transfer all the material of the eaten into the tissues of the eater. It is an inefficient process that, at most, makes use of only about 10 percent of the mass of the eaten to build up the tissues of the eater. The other 90 percent is discarded as waste or is converted into energy that powers the activity of the eater's body and is then given off as heat.

Therefore, if only animals existed, even though they

were of many different species, they would soon eat each other into extinction.

In the world about us, most animals live on plants. Some animals live on other animals, but then the animals that have been eaten are likely to have lived on plants. Even if animals eat other animals who eat other animals and so on, the whole chain, in the long run, rests finally on an animal that eats plants. That enables animals to exist indefinitely.

How can that be? Don't plants have to eat also? Don't they have to gain the energy with which to keep their tissues alive just as animals do?

Yes, but in the case of plants, the food is not the tissues of any other living thing. The food is carbon dioxide from the air, plus water and minerals from the ocean or soil, and the energy supply is absorbed from something as simple and as seemingly endless as sunlight. Given simple molecules and sunlight, plants can grow and multiply indefinitely despite the depredations made upon them by animals who continually raid the food that plants so painstakingly build up.

Plants can make use of sunlight because of a green chemical, *chlorophyll* (from Greek words meaning "green leaf"), that they contain and animals do not. Therefore, when we talk of plants utilizing sunlight, we mean green plants and not those plants without chlorophyll, like mushrooms.

This means that the complex animals of today could not live in the sea unless plants also existed there and had developed first. Furthermore, animals could not have invaded the land, unless plants had also done so and done it first.

Plants that live in the sea have always been of very simple structure and still are today. They float in the uppermost layers of the sea, where they can receive the sunlight they require as an energy source. (Sunlight is totally absorbed by the top 250 feet [75 meters] of water, so that in deeper water plants do not live. Animals can, of course, penetrate to any depth.)

These simple plants of the sea absorb water, minerals, and even carbon dioxide directly from the sea around them, and back into the sea they can discharge their wastes (including oxygen, something about which we will have more to say later on in the book). These simple plants are, for the most part, microscopic blobs of life called *algae*, and the most complex forms, such as seaweed, are merely conglomerations of algae. (Indeed, *algae* is Greek for "seaweed.")

In order for plants to live successfully on land, they must have some waterproof outer surface that will keep them from drying out in the largely waterless surroundings. They must also have some stiffening agent that will allow them to grow upright despite the pull of gravity and to spread parts of themselves outward to catch the sunlight they require. They must develop roots that will hold them firmly in the ground and that will absorb water and dissolved minerals from the soil. They will also have to have a system of ducts that will conduct the water and minerals from the roots to all parts of the plants.

Land plants are much more complex than sea plants, and the difference between the two is far greater than between land vertebrates and sea vertebrates, or between land arthropods, mollusks, and worms and sea arthropods, mollusks, and worms.

If the amount of change that was required constituted the only criterion, we would expect that plants adapted to the land long after animals had done so.

However, no matter how comparatively easy the transition would have been for any form of animal life, that transition would have to wait for plants to be the first successful invaders. Plants had to make it to land first so as to serve as a source of food for animals, before animals could also make the move.

Plants made the advance before the opening of the Devonian. They reached land during the Silurian period, which extended from 410 MYA back to 440 MYA, a duration of 30 million years.

The first known plants capable of living on land had no

roots and consisted of a simple forked stem without leaves. They did, however, possess a vascular system—that is, ducts for the transmission of water and dissolved materials. They made their timid appearance at the edge of the shore some 450 million years ago.

This would mean that plants had 50 million years to multiply and diversify in a peaceful Eden free of animal life. (To be sure, plants compete silently but fiercely among themselves—for ground water, by elaborating competing root systems, and for light, by climbing high and spreading wide.)

By the time the mainstream of animal life—insects and amphibia—had ventured out onto land, the plant world, before the Devonian was over, had sprouted and expanded into tall trees and formed the first forests.

But now let us get back to the amphibia. They did not spring from nowhere but evolved from fish that were sea vertebrates. What was the beginning of the vertebrates? Or since vertebrates are part of the phylum Chordata, which includes a few related invertebrates, what was the beginning of the chordates?

# 12

## CHORDATES

In the Devonian period, when the land was turning green and animal life was venturing onto land, the seas were rich with fish. In fact, it is the Devonian period that is sometimes specifically called the Age of Fish.

Fish are still the dominant form of life in the sea today, 350 million years after the close of the Devonian. Now, however, there are land chordates that have returned to the sea to a lesser and greater extent (sea snakes, sea turtles, penguins, seals, manatees, dolphins, whales, and so on) that compete with fish in their own element and prey on them. There are also land animals that, though not really sea creatures, feed on fish to a considerable extent, as, for instance, herons and otters. In the Devonian, there was no such competition or danger, for reptiles, birds, and mammals did not yet exist.

The most successful group of fish, at the present time, are the *Actinopterygii* (Greek for "ray fins," because the fins consist of skin stiffened by horny rays). Ray fins are excellent for paddling.

The first ray-finned fish appeared about 390 million

years ago in the early Devonian and now make up by far the major portion of the fish species. Like all sea creatures, they can grow to large size. The largest modern ray-finned fish is the sunfish, an occasional specimen of which may be over 2 tons in weight.

In the Devonian, a second group of fish, the *Sarcopterygii* ("flesh fins") were as successful as the ray fins, if not more so. In the flesh-finned fish, the fins consisted of a lobe of flesh and bone, fringed with the skin and rays of an ordinary fin.

The flesh-finned fish were less adept at paddling, but they could support themselves on their fins as ray fins could not. The flesh fins could maneuver about at the bottom of the sea and, if they lived in shallow water, could eventually clamber out on land for periods of time.

It may be that when the ray fins and the flesh fins first appeared, they were shallow-water creatures that developed simple sacs into which they could gulp air, from which they could absorb oxygen. Such sacs supplement the action of gills and help out if the shallow water turns brackish and muddy. These sacs were primitive lungs.

The ray fins, with their excellent system of paddling, could move into deep water, where the gills worked adequately and well. They did not need the primitive lung, and it became an air sac that by containing less or more air, could make them less or more buoyant and helped them sink or rise in the water.

The flesh fins, on the other hand, tended to keep the lungs, at least in some cases. After the Devonian, however, the flesh fins, with their more limited way of life, began to lose ground to the ray fins, who could exploit the whole ocean. During the Mesozoic, the flesh fins dwindled, and today very few of them survive.

There are a few species of *lungfish* that survive even today. They live in restricted areas in Australia, central Africa, and central South America, always in areas subject to drought so that there is an advantage in being able to gulp air. Some lungfish can even survive conditions where the water they live in dries out completely. They then remain caked in dry mud, in a kind of *estivation*, which is the summer equivalent of the more familiar winter hibernation. When the rains come, the mud softens, pools form, and the lungfish swim away.

One might think that the lungfish, with their lungs, were the ancestors of the amphibians, and from them all the rest of the land chordates evolved, including us. That would be wrong, however, for lungfish have certain characteristics that one doesn't find in the early amphibians so that the former are not likely to be ancestral to the latter.

Another group of flesh fins are the *crossopterygians* (from Greek words meaning "fringe fins"). The bones in their fins had the basic arrangement of bones in the early amphibians (and of the bones in our own limbs, for that matter). In various other ways, too, they resembled the later amphibians.

It is thought that a particular type of crossopterygian fish called *Rhipidistians* ("fan sails") gave rise to the amphibians, and then died out about the time, or just before, the Permian extinction. The modified Rhipidistians—the amphibia—survived the extinction and went on to further evolution.

Indeed, it was thought for a long time that all the crossopterygians had become extinct about 150 million years ago, toward the end of the Jurassic, at a time when the dinosaurs were flourishing.

Then, on December 25, 1938, a trawler, fishing off South Africa, brought up an odd fish about 5 feet long. A South African zoologist, J. L. B. Smith, who had the chance of examining it, recognized it as a matchless Christmas present, for it was clearly a crossopterygian.

It was not, of course, a Rhipidistian; those *are* extinct, as far as we know. What had happened was that although the crossopterygians were primarily freshwater fish (and amphibians are freshwater animals to this day), one branch evolved the capacity to live in salt water and moved into the ocean. These were the *coelacanths* ("hollow spines," one of their features). The coelacanths lived at fairly deep levels in the ocean and escaped notice until 1938.

Smith had first received notice of this strange fish from a Miss Latimer at a local museum to whom the fisherman had brought his specimen. Smith therefore named this species of coelacanth "Latimeria" in her honor.

Latimeria is not our fish ancestor, of course, but it is the only living crossopterygian, as far as we know, and we are descended from another sort of crossopterygian.

The ray fins and the flesh fins, taken together, are the *Osteichthyes* ("bony fish"). They resemble each other in having a well-developed bony skeleton featuring a spinal column of vertebrae.

The oldest of the bony fish may date back to the beginning of the Silurian period, about 440 million years ago. These were not the first organisms to have an internal skeleton, but they were the first to have one made of bone. This took place as the plants were just venturing out on land, and when no animals had yet done so. Internal bones are thus older than animal land life.

Yet bone need not be inside the body. During the Devonian, there were fish that were not osteichthians. They were the *placoderms* (Greek for "plated skin"). They had internal skeletons of cartilage, which was made up of tough protein fibers but lacked the mineral content, mainly calcium hydroxyphosphate, that bone has. (You can feel the difference in your nose. The tip is stiffened with cartilage, which is flexible and can bend. The upper portion is stiffened with bone which is hard and unyielding.)

Nevertheless, the placoderms possessed bone in the form of armor around their head and the forward part of their trunk. It was this outer bone that made the "plates" that gave them their name. This external bone served as an armor that protected them against predators. Such protection seems like a good thing, but it comes at a price. The armor, to be effective, must be strong and, therefore, thick and heavy. The placoderms were poor swimmers in consequence and tended to be bottom dwellers.

In general, among animal life forms, mobility seems more successful than armor. Thus, among the mollusks, squid seem to do better than oysters; among reptiles, lizards do better than turtles; among mammals, rodents do better than armadillos.

The placoderms seem to bear out this notion, for although they were very common in the Devonian and some were fearsome creatures who were up to 30 feet (9 meters) long, they did not make it. By the end of the Devonian, they were almost all gone.

Or, rather, the external armor was all gone. The bony plates grew thinner, for the thinner the plates, the faster and more efficient the swimming, and the advantage of this made up for the weakening of armor. Eventually, there were placoderms with no armor at all, and from these the modern sharks and related species have probably descended, making their first appearance about 390 million years ago.

Sharks are not bony fish. They differ from the bony fish in the position of the mouth, the lack of a gill plate covering the gills, and their asymmetric tails. The most important difference to zoologists, however, is that sharks and related animals do not have bones. They have an internal skeleton, to be sure, but it is entirely made up of cartilage. The sharks and related species are, therefore, the *Chondrichthyes* (Greek for "cartilage fish").

Sharks are not particularly handicapped by this. Cartilage is not as strong as bone and it wouldn't do for land life. When an animal is as large as a brachiosaur, an elephant, or even a man, nothing but bone will do to withstand gravity. That is why it was bony fish that emerged on land. No shark ever did so. They remain today, as they were at the start, exclusively animals of the water.

In the water, however, cartilage is quite strong enough to support the body. In fact, since cartilage is lighter and more flexible than bone, it makes for better swimming. Certainly, sharks are efficient swimmers and fearsome predators. The great white shark, which is the largest of the carnivorous sharks, can be 15 feet (4.5 meters) long and weigh well over a ton. It was this shark that was the frightening monster of the movie *Jaws*.

There are larger sharks, too, but these do not feed on large prey but filter out the tiny plants and animals that float in the sea (as the largest whales do). There is much more of these small organisms than of large, and the former can support larger animals. The largest shark is the *whale shark,* some of which can approach 60 feet (18 meters) in length and have a weight of more than 40 tons. There may have been now-extinct sharks that approached 80 feet (24 meters) in length and rivaled the largest whales in size.

The sharks and the bony fish have a number of features in common. They both have an internal skeleton, whether cartilage or bone. They both have two pairs of fins, which set the fashion for the four limbs of all the later chordates, including ourselves.

(Of course, in some cases, such limbs atrophied and disappeared, the hind two in whales, the front two in kiwis, all four in snakes, but no chordate has ever had a true fifth limb. There are some animals, notably spider monkeys and opossums, that have a tail that is prehensile and serves almost as a fifth limb of sorts—to say nothing of an elephant's trunk.)

Then, too, and perhaps most important of all, the sharks and the bony fish both have jaws. A primitive gill arch was bent in the middle and became capable of opening and closing. Equip the opening with hard teeth, and you have a very effective weapon and tool.

The cartilaginous fish and the bony fish can therefore be lumped together as the *jawed fish,* and the first jawed fish may have been a primitive placoderm that dated back about 450 million years ago, in the Ordovician period that preceded the Silurian. The Ordovician period dates from 440 MYA to 500 MYA, a duration of 60 million years.

Yet there are still more primitive fish, fish that lack jaws, called *agnathous* (Greek for "no jaws"). In the Devonian, when a wide variety of fish of all kinds lived, there were the agnathic *ostracoderms* (Greek for "shell skins") which, like the placoderms, had external bony armor but lacked jaws and had not developed two pairs of fins. Most were probably bottom-dwelling organisms that sucked water into their forever-open mouths and filtered out anything, living or dead, that could be digested.

The ostracoderms were no more successful than the placoderms in their competition with mobile, unarmored fish. By the end of the Devonian, they were gone, leaving behind unarmored descendants, a few of which still persist today. The best-known present-day agnath is the lamprey, which looks like an eel but has no paired fins, no scales, and, of course, no jaw.

The ostracoderms were the first organisms to develop bone, but, like the placoderms, their internal skeleton was cartilaginous. They also had a spinal column made up of vertebrae.

What all these various fish—with and without jaws, with and without paired fins, with and without bones— have in common are the internal skeleton, and the spinal column made up of vertebrae. All the descendants of these fish that ventured out on land and evolved further there— amphibia, reptiles, birds, and mammals—also possess this internal skeleton and the spinal column made up of vertebrae.

They are all, from agnaths to man, classified therefore as *vertebrates*. The earliest vertebrates were the ostracoderms that may have first appeared at the beginning of the Ordovician about 500 million years ago. If, then, you feel the knobs that run down the middle of your back, you are feeling a body feature that is half a billion years old. The bone of which those knobs are formed, though not always present in the vertebrae, has been around for half a billion years also.

Yet all the vertebrates belong to the phylum Chordata. Are the vertebrates all there are to the chordates? Or are there chordates that are *not* vertebrates?

Here is the way we can reason it out. All vertebrates have a central nerve cord that is hollow and runs along the back. The nerve cord is, indeed, enclosed by the spinal vertebrae. In all other phyla, the nerve cord, if it exists, is solid, not hollow, and runs along the abdomen, rather than the back.

Secondly, all vertebrates have throats that are perforated

by gill slits through which water can be passed. From that water, food can be filtered out and oxygen can be absorbed. These are not present in other phyla. To be sure, in land vertebrates like ourselves such gill slits do not exist, but if we follow the embryonic development of such vertebrates, we find that at an early stage gill slits begin to develop, but that they wither away. This is true even in the human embryo. There are many such traces of more primitive stages in embryonic development—the human embryo has the beginnings of a tail for a time, as an example. Such things are among the many strong lines of evidence in favor of biological evolution.

Thirdly, all vertebrates have, at some time during their embryonic development, an internal stiffening rod of a tough, light, flexible, gelatinous substance that runs down the back. This is called a *notochord* (Greek for "back string"). In all vertebrates this is replaced by vertebrae before the embryonic development is completed, but it is always there at first.

Suppose there are living organisms that have a hollow, dorsal nerve cord, gill slits, and a notochord. They should therefore be considered as related to vertebrates even though they never develop vertebrae or any of the other specialized characteristics of the agnaths and their descendants.

These nonvertebrates would be lumped together with the vertebrates to form the phylum Chordata (named for the notochord). Actually, such nonvertebrate chordates do exist, although there are only a few of them and none are markedly successful members of the family of life.

There is, for instance, a little fishlike organism, no more than 3 inches (7.5 centimeters) long at most. It has no distinct head but is pointed at both ends so that it looks rather the same coming or going. It is called *amphioxus* (Greek for "both pointed").

It is an extremely primitive creature, without even a brain, but it does have a hollow dorsal nerve cord, it does have gill slits, and it does have a notochord running the length of its body. It has no internal skeleton except for the notochord, and it has no vertebrae, so it is not a vertebrate, but it is, nevertheless, a chordate.

Then there is the *tunicate,* which is as motionless as an oyster, although rather than a shell, it has a tough, leathery outer tunic (which gives it its name). It has no notochord, it has no nerve cord—but it *does* have gill slits, lots of them.

The tunicate mentioned here is, however, the adult form. When the tunicate eggs hatch, the result is a larval form that is to the adult tunicate what a tadpole is to a frog. Indeed, the tunicate larva looks rather like a tadpole. It has a head with gill slits and a tail that enables it to move about. In this tail is a long notochord and, above it, a dorsal nerve cord. When the tunicate settles down to its adult life, it absorbs the tail, which disappears along with the notochord and nerve cord it contains, but the tunicate is a chordate just the same.

Finally, there is an organism that looks rather like a worm. At the front end, there is a proboscis shaped vaguely like a tongue or an acorn, so that it is sometimes called an "acorn worm." Behind it is a collarlike structure like a barnacle, so that it is also called *balanoglossus* (Greek for "barnacle tongue"). Behind the collar there is a long wormlike extension, but just behind the collar in the front part of this extension there are gill slits. What's more, in the collar there is a remnant of a dorsal nerve cord and, sticking into the proboscis, a small piece of notochord. The balanoglossus is also a chordate.

It seems inevitable, then, that the ancestors of the first agnaths were simple invertebrate chordates. There are no fossil remnants of such organisms, but we can guess that the beginning of the chordates came about 550 million years ago. This is in the Cambrian period, which extends from 500 MYA back to 600 MYA, a duration of 100 million years.

If so, the chordates are the latest of the phyla to be established. In the Cambrian, all the other phyla, as nearly as we can make out, were already well evolved and flourishing. It would seem, then, that having penetrated back to the beginning of the chordates, we ought to strive to go further back still and grope for the beginning of life itself.

Before we do that, however, we might want to stop and ask how far back we can possibly go. We have gone back over half a billion years, and life is still flourishing and various. Does our Earth, the stage on which life exists, go much further back? How old is the Earth?

Or, since that is an enormous question perhaps best tackled in stages, let us ask first how old the land-sea configuration that we are accustomed to might be. In other words, what is the beginning of the continents?

# 13

CONTINENTS

It is quite obvious to anyone who thinks of it that human devices are developed through a process of evolution. It is scarcely credible that anyone would argue with that.

It has become plain, too, that life itself develops through a process of evolution. This is not quite obvious to most people, and there are strong emotional (not rational) reasons that cause many people to doubt it. Nevertheless, biological evolution is accepted by scientists, who consider the fact beyond dispute even though the details of the process elicit considerable argument.

It is, however, tempting to look upon the Earth as being beyond evolution. One might suppose it just remains as is, the passive stage on which the events of life, human and nonhuman, take place. Granted, hills may be leveled, canals may be dug, marshes may be filled in, and rivers dammed or diverted by human effort, but these are relatively small things, and if we subtract human effort we might certainly suppose there is no essential change on Earth.

Thus, we say "as old as the hills" when we mean

"indefinitely old," for the hills have surely always been there. Alfred, Lord Tennyson (1809–1892), speaking of an insignificant little brook, wrote the famous lines: "For men may come and men may go,/ But I go on forever." Surely we can guess that brooks are not for eternity but can come and go with relatively slight changes in the environment, but emotionally we accept the permanence of inanimate objects. Even the Bible says in Ecclesiastes 1:4, "One generation passeth away, and another generation cometh: but the earth abideth forever."

Things inanimate do seem to be permanent in terms of human lifetimes, and yet people really hesitate to think about things in terms of "forever." Eternity is a difficult concept and doesn't seem to fit with what we know. All living things have a beginning, for all are born at some fixed time. All human devices have a beginning, for all are constructed at some fixed time. Might it not be, then, that even the Earth follows what might seem to be a universal rule, and might it not have been constructed at some fixed time?

Naturally, the Earth, being far beyond anything of human origin in terms of size, grandeur, and complexity, requires a constructor or "Creator," equally beyond the human in size, grandeur, and complexity. Earth must therefore have been created by superhuman beings, which we can refer to out of long habit as "gods."

Thus, the Babylonians felt that in the beginning there was Tiamat, who represented the boundless waste of salt water, or chaos. (Matter apparently always existed, but what did not always exist was order and organization. It was that which had to be created.)

Out of chaos, somehow, gods and goddesses were born, representing organizing principles. The tales of these early gods are confusing because every city-state in the Tigris-Euphrates valley had its own gods, and their adventures and misadventures may well have reflected the ups and downs of their respective city-states in the perennial wars that occupied them.

Eventually, Marduk was recognized as the chief god,

the main organizing principle. Why not, since he was the god honored at Babylon, which, about 1725 B.C., became the dominating city of the lower valley and remained so for fourteen centuries. The gods battled Tiamat, and Marduk slew her, thus establishing the principle of order.

Marduk then proceeded to impose order on chaos by making use of the monstrous body of Tiamat to establish a Cosmos (the opposite of Chaos; ordered matter rather than disordered matter).

He split Tiamat's body and made the sky out of one half, the Earth out of the other. Various other parts of the body became earthly phenomena—her blood became the seas, her bones became the rocks of the dry land, and so on.

Undoubtedly all this could be interpreted allegorically by philosophers and could be made to end up as a respectable cosmogony, considering the amount of data then available. However, the general population undoubtedly accepted the tale as literally true, and any attempt to deviate from it would be considered blasphemous and dangerous.

The Jews who were in captivity in Babylon in the sixth century B.C. picked up the Babylonian tales of the creation and adapted it to their own use. The Jewish leaders had no use for anthropomorphic gods (at least by that time) and did not want to imagine God fighting the monster of Chaos, though there are passages in the Bible that indicate in poetic fashion that that is just what the old myths had him doing.

Instead God did not arise out of chaos but existed eternally. He "moved upon the face of the waters" (Genesis 1:2)—the original Chaos. God then performed the creation in steps, but did so by his mere word. His will alone imposed order. The tale is really powerfully poetic and far advanced beyond any creation tale invented earlier.

The creation tale of Genesis is very impressive, even in modern terms, if it is treated symbolically and allegorically. But again, the tendency for many people is to accept it literally and to fight ferociously against deviating from it by one iota.

The same pattern of the creation by supernatural gods of an ordered Universe out of Chaos occurs over and over again in various mythologies, and in a sense, that is the only story possible. Even modern scientists are forced to work out methods whereby an ordered Earth can be created out of original chaos, but to bring this about they cannot use gods who work with foresight and will, after the human fashion, but must make use of the ineluctable laws of nature that act out of necessity and without deviation.

This is by far the more difficult task and depends upon evidence and reasoning therefrom, rather than upon romantic and poetic imagination. That is why the scientific version of the creation of Earth did not come until thousands of years after the various mythological versions had been established.

It is easy to believe that the Earth, having been created by God, would be created as the perfect abode for life (especially human life) from the very beginning and, therefore, would not change (except by the direct will of God, as in the Flood, the destruction of Sodom, the parting of the Red Sea, and so on). To suppose it to change otherwise, would be to accuse God of producing something imperfect or of imagining his creation able to undergo further changes on its own without his help.

And yet changes without human intervention are noted. Brooks *do* dry up, rivers change their courses, other rivers build deltas out into the sea using the silt that they carry along with them. The shoreline undergoes slight changes this way and that, fissures form in the Earth due to earthquakes, volcanoes come to life. All these things, however, can rightfully be dismissed as minor and even trivial.

None of the mythological versions of the beginning of Earth place a date on that beginning, even an approximate one. All the accounts, even the biblical one, might just as well start "Once upon a time . . ."

As we said earlier, Archbishop Ussher calculated that the Earth was created in 4004 B.C., but it is he who said so, and not the Bible. Still because that date (or something like it) was so largely accepted, it was an enormous stroke in favor of the changelessness of the Earth. From the observed rate at which changes take place, the total change that could possibly take place in 6,000 years would be entirely insignificant.

Of course, even the ancient Greeks noticed certain things that spoke of big changes, not small ones. For instance, the ancient philosophers noted the presence in mountainous areas of rocky remnants of what were clearly seashells. They were forced to speculate that what were now mountaintops had once been under the sea. Since the ground level was not noticeably changing, those mountaintops could only have been under the sea ages ago, for if that portion of the land surface was rising, it was at a rate too slow to measure in a human lifetime. Other thinkers in later centuries kept noticing this over and over and coming to the same conclusion.

To such speculations, however, the biblical scholars had a ready answer. It was the tale of Noah's Flood, which, according to the Bible, had been worldwide and had covered even the highest mountains. Naturally, such a flood would wash seashells onto mountain peaks. Indeed, the cataclysm of a universal Flood could be called upon to account for any drastic geological change for which evidence seemed to exist.

Except for the biblical Flood, the most notorious example of an ancient tale of large changes in the Earth was the account, by the Greek philosopher Plato (427–347 B.C.), of the sinking of Atlantis. A whole continent located beyond the Strait of Gibraltar, in the then-misty and unknown Atlantic Ocean, had sunk beneath the waves in a single day as a result of an earthquake. Plato placed the date as 9,000 years before his time, or 9400 B.C.

Of course, the account could be accepted as a fable, as a piece of fiction designed to make a point. However, even fables are often based on some half-remembered, time-distorted fragment of history, so that it could be easily maintained, for instance, that Plato was recounting a dim memory of the Flood that, at least temporarily, drowned all the continents. Or, if there was actually a continent beyond Gibraltar that sank, that might have been the work of the Flood.

In actual fact, Plato's Atlantis story is now thought to be based on an event more recent than the date usually ascribed to the Flood.

Eleven centuries before Plato's time, the island of Thera in the Aegean Sea, about 150 miles (240 kilometers) southeast of Athens, had a flourishing civilization related to that of the larger island of Crete, 80 miles (125 kilometers) still farther south.

Thera, however, was not merely an island, but was the tip of a volcano jutting above the sea. About 1500 B.C., it erupted in a vast explosion that destroyed the island in a very brief period and let the sea roll over the remnants. The explosion, the showers of ash, and the tidal waves that roared over all the neighboring shorelines devastated Crete and perhaps may have helped give rise to a Greek flood myth, independent of Noah's.

The event was never quite forgotten but was, of course, exaggerated and distorted over time. Naturally, it was made more romantic by being placed in the remote past. In the memory of later humanity, the Earth showed no such convulsions—an occasional volcanic eruption, an occasional earthquake, yes, but these were clearly local phenomena.

Then, in 1492, there came the discovery of the American continents by the Italian explorer (in Spanish pay) Christopher Columbus (1451–1506), even while Portuguese explorers were working their way around the tip of Africa in order to reach India.

In the following century, the new shorelines of South America and Africa were mapped. As a result, a rather astonishing thought forced itself on those who looked at the new maps. The first one we know of who put the thought into words was the English philosopher Francis Bacon (1561–1626). In 1620, in his book *Novum Organum,* he mentioned that the eastern coast of South America almost exactly matched the western coast of Africa, so that they would nearly fit if they were imagined to be pushed together. This, he maintained, could not be mere coincidence.

The implication was, of course, that South America and Africa had once been together and had somehow been pulled apart. But how could that have been done?

However, almost as soon as Bacon had made his observation, it was pointed out by the traditionalists that Africa and South America, if they had indeed once been joined, could easily have been torn apart by the mighty chaotic force of the Flood.

By then, though, the Flood was being questioned by some brave men. About 1570, a French potter (and thinker), Bernard Palissy (1510–1589), pointed out that nature was changing the land even as men watched. Rain, together with the battering of wind and waves, was wearing down the mountains and eroding the shores. He maintained that this was enough to cause great changes without any necessity of supposing there had been a universal Flood. He also thought that fossils were the remains of once-living animals.

These were dangerous times for men with unpopular views, however. The Protestant Reformation had begun in 1517, and all western Europe was taking sides in a confrontation between Catholics and Protestants that was to result in over a century of religious wars. Palissy was a Protestant in a France that was chiefly Catholic, and both sides were particularly sensitive to the dangers of dissidence and any questioning of accepted religion. Palissy was accused of heresy, condemned, and burned at the stake in 1589. Undoubtedly, his denial of the Flood was a serious piece of evidence against him.

Eleven years later, in 1600, the Italian philosopher Giordano Bruno (1548–1600) was burned in Rome for heresies that included his belief that the Earth turned about the Sun, that the stars were other suns with other worlds revolving about them, and so on. And in 1633, the Italian scientist Galileo Galilei (1564–1642), was threatened with torture by the Inquisition and was forced into a public admission that he was wrong in believing the Earth moved around the Sun.

Scientists were forced to be cautious. In 1634, the French philosopher Rene Descartes (1596–1650) heard what had happened to Galileo and abandoned his own plan to publish a book in which he intended to describe the formation of the Earth according to natural processes. He felt it wouldn't be safe to do so, and one can scarcely blame him.

The Danish geologist Nicolaus Steno (1638–1686), like Palissy, believed fossils were the remains of once living animals. In 1669, however, he advanced tortured explanations that served to adjust his beliefs to biblical legends.

As late as 1681, an English clergyman, Thomas Burnet (1635–1715), wrote a book that supported the story of the Flood on geological principles (as he understood them) and concluded that the Earth had remained unchanged since the Flood and would, presumably, continue unchanged until it was God's will to destroy it. Nevertheless, in 1691 he wrote another book in which he refused to accept the Adam and Eve story as literal truth, but only as allegory. This got him into trouble. He wasn't exactly mistreated, but he was denied all further promotion.

Even the sternest oppression, however, cannot stop human thought forever. Not even the threat of punishment in life or hellfire after death can stop people from observing, thinking, and reasoning.

A French naturalist, Georges Louis de Buffon (1707–1788), began, in 1749, a long encyclopedia of natural history that eventually ran to forty-four volumes. In it, he did what Descartes had feared to do a century earlier and attempted to explain the Earth's origin in completely naturalistic terms.

In the very first volume, he suggested that the Earth might have been created by the catastrophic collision of a massive body with the Sun, and that the Moon was somehow then ripped from the Earth. The Earth gradually cooled down, water vapor condensing as it did so and forming the oceans. Fissures appeared in the Earth through which much of the water drained, exposing the continents.

All this took time, and Buffon estimated that the Earth

had already been in existence for 75,000 years, that it was continuing to cool, and that it might last an additional 93,000 years before becoming too cold to be habitable. Life had begun on Earth, he estimated, about 40,000 years before his time.

We now know that Buffon's estimate of the time scale was immensely short of the truth, but this was the first serious and public attempt to go beyond Archbishop Ussher's limits. Naturally, he got into trouble with these views and was forced eventually, like Galileo, to recant them and to proclaim publicly that he was in error.

In not all nations, however, were religious forces so powerful as to be able to punish independent scientific thought. In Great Britain and in the newborn nation of the United States of America, religious organizations might denounce any attempt at thought but could not rally actual force against dissidents.

Thus, in the United States, the American statesman Benjamin Franklin (1706–1790), who was far ahead of his time in almost every way, suggested in 1784 that the Earth's crust must be a relatively thin shell floating on hot fluid and that this shell could break up and slowly shift around, producing vast changes as it did so. It was a remarkable thought, and it took nearly two centuries for the rest of the world to catch up.

Franklin's thought, however, was just a speculation thrown out for the world to ponder. It lacked any carefully worked-out series of observations to back it, so that it must have seemed merely an odd and interesting fancy to those who heard of it.

The real breakthrough came in Great Britain.

A Scotsman, James Hutton (1726–1797), grew prosperous as a chemist, sufficiently prosperous to retire in 1768 and devote himself to his hobby, which was geology. (In fact, he is sometimes called the "Father of Geology.")

His observations led him to Palissy's conclusion that there were natural processes affecting the Earth and bringing about a slow evolution of its surface structure. Some rocks, it seemed clear to Hutton, were laid down as sediment and compressed to eventual hardness; other rocks were molten in the Earth's interior and were then brought to the surface by volcanic action. Exposed rocks of either type were worn down by wind and water.

His great intuitive addition to all this was the suggestion that the forces now slowly operating to change the Earth's surface had been operating in the same way and at the same rate through all Earth's past. That is the *uniformitarian principle,* which was opposed to the *catastrophism* of men such as Bonnet.

Judging by the slow rate at which sedimentation, volcanic action, and erosion worked and, on the other hand, by the thick layers of sedimentary rock laid down and the vast river deltas that formed, Hutton had to conclude that Earth must have been in existence a very long time. He said, in fact, that he could find no vestige of a beginning and no prospect of an end. This did not mean that he thought Earth was eternal, merely that both its beginning and its end were so far off that he could see no evidence that would lead him to measure either time in reasonable fashion, and in this he was correct.

He published a book, *Theory of the Earth,* in 1785, in which he presented his views. Unlike so many of his predecessors in dissent, he was not persecuted as a result, but neither was he rewarded. The weight of theological disapproval was heavy, and since the book itself was not an easy one to read, it might have seemed at first that it would have little influence on scientific thought.

Some scholars, however, did read it and were impressed. Another Scottish geologist, John Playfair (1748–1819), published a book in 1805 (after Hutton's death) in which he explained Hutton's theories in a sprightlier and more popular form, and after that those ideas began to spread more rapidly. Since Hutton's reasoning made it possible to think of a truly long-lived Earth, scholars began for the first time to think of Earth's existence not as a matter of thousands of years as Ussher did, or even of tens of thousands of years as Buffon did, but as a matter of *millions* of years.

The German naturalist Friedrich Wilhelm Humboldt (1769–1859), who explored South America between 1799 and 1804, took up Francis Bacon's old observation of the similarity of South America's eastern shore and Africa's western shore. He showed that the similarity lay not only in the shapes visible on the map, but in geological similarities as well. These all but demanded that the two continents be considered as having been once joined. However, Humboldt did not live in Great Britain or the United States, and he did not quite have the courage to explain the matter on Huttonian principles. He fell back on the Flood.

The French naturalist Jean de Monet de Lamarck (1744–1829), in 1809, was the first to describe a possible mechanism for biological evolution. The mechanism was wrong, and the world had to wait half a century more for Charles Darwin to advance the right one. Still, Lamarck's theory was the first to take advantage of the notion of the long-lived Earth. It began to persuade people that although evolution proceeded at so slowly a rate as to be impossible on a short-lived Earth, it became a practical possibility in a long-lived one.

It was yet another Scottish geologist, Charles Lyell (1797–1875), however, who put Huttonian ideas over the top. Between 1830 and 1833, he published a three-volume work entitled *The Principles of Geology*, in which he organized and explained Hutton's theories, together with observations and advances made in the half century since Hutton had published his book. Lyell's book proved utterly convincing and it impressed, among others, the young Charles Darwin, who began thinking of biological evolution as a result. Since Lyell's book, no scientist has seriously doubted that the Earth was long-lived.

Lyell named some of the geological periods I have been referring to in this book: the Eocene, Miocene, and Pliocene, for instance. He also made an estimate of the age of the oldest fossil-bearing rocks, suggesting that they were 240 million years old. This was the first time the possibility was raised that the Earth might not merely be millions of years old, but hundreds of millions of years old.

However, it is not the age of the Earth itself with which we are concerned in this section, but only with the nature and time scale of the changes of the surface features.

The long-lived Earth was adopted, perforce, even by those geologists who were deeply religious. One such was an American geologist, James Dwight Dana (1813–1895). (Late in life, he even reluctantly accepted Darwin's mechanism of biological evolution.)

About 1850, Dana returned to Buffon's notion of a cooling Earth, and he could now imagine it cooling very slowly over long periods of time. It seemed to him that as it cooled, the crust solidified. Some parts of the surface, for some reason, solidified first, and these parts were the continents as we know them today. (Dana clearly thought the continents existed in their present positions and shapes from very nearly the beginning.)

As the Earth continued to cool, it shrank (as most cooling objects do). The already solidified continents resisted change, but the still-fluid areas between responded to the shrinkage by sucking inward. Thus were formed the sea bottoms. As it cooled, the water vapor surrounding the Earth condensed and formed liquid water. This water, coming down in an all but endless rain, collected in the ocean basins, forming the land-sea pattern that still exists today.

As the Earth continued to cool still further, it shrank still further, and finally the continents were forced to accommodate themselves to the slightly smaller Earth by wrinkling—the wrinkles being the mountains.

The theory was very impressive, but it had some holes in it. Why did some of the surface solidify early to form the continents? Why did the mountains form only in certain regions of the continents instead of all over? It was also clear that mountains formed during relatively brief mountain-building periods separated by relatively long intervals, and that during much of Earth's history, there was no mountain building. Why was that?

There was another problem. Studies of natural history and of fossils showed that similar plants and animals existed in widely different parts of the world. Similar species of plants and animals of types that couldn't possibly have crossed wide ocean barriers existed in both Africa and South America, or in both India and Australia. Then, too, the island of Madagascar, off the east coast of Africa, had few species in common with Africa but many species in common with India, which was much farther away.

Since Dana's notion that the continents did not change their positions was accepted at the time, it seemed necessary to suppose that in the past there were "land bridges" between different parts of the Earth's land areas, where ocean water now rolled.

An English naturalist, Philip Lutley Sclater (1829–1913), suggested in 1864 that there had been a land bridge between Madagascar and India at one time. As the earth cooled and shrank, the land bridge broke up and collapsed, sinking beneath the sea. Before it did so, however, life had spread freely between India and Madagascar. Since numerous species of lemurs are to be found in Madagascar, the land bridge was named *Lemuria*.

The notion of land bridges reached its peak with the Austrian geologist Eduard Suess (1831–1914). He wrote a three-volume work called *The Face of the Earth,* which was completed in 1909. In order to explain the distribution of life, he imagined that there had at one time been a massive supercontinent, which he called *Gondwanaland* (after a portion of India which was included in it.) This supercontinent consisted of South America, Africa, India, Australia, and Antarctica, with land bridges in between. There were other continents in the north, with the "Tethys Sea," a kind of precursor of the Mediterranean Sea, in between.

Actually, the whole notion of a shrinking Earth, of wrinkling mountains, and of land bridges was wrong. However, the geologists, from Dana to Suess, had succeeded in planting the idea that the Earth's crust had undergone evolutionary changes. It remained to find out what the correct changes were.

The land-bridge theory involved the notion that the land

masses stayed where they were, but moved up and down. Was it possible that they moved sideways?

As early as 1858, an American, Antonio Snider-Pellegrini, had written a book in which he suggested that as the Earth cooled, one large continental mass formed on one side of the world. Somehow it broke up and pulled apart into the continental arrangement of today. But how did it break up and pull apart?

Snider-Pellegrini suggested it was the action of Noah's Flood, which at once killed the idea. In the post-Lyell period, no serious scientist would accept the Flood as the causative agent of anything. Nevertheless, if some other way of accounting for such a sidewise motion arose, the Snider-Pellegrini idea might not seem so bad.

Even earlier, in 1735, a French scientist, Pierre Bouguer (1698–1758), was exploring the Andes Mountains in South America and was calculating their height. He tried to establish a vertical line by suspending a heavy weight from a support. He expected that it would be diverted slightly from the exact vertical by the gravitational pull sideways of the neighboring mass of lofty mountains. The diversion was substantially less than he had expected, which meant that the mountains were less massive than they seemed.

A hundred years later, an English surveyor, George Everest (1790–1866), after whom the highest mountain in the world, Mt. Everest, is named, was surveying the Himalayas and obtained similar results. Those mountains weren't as massive as they looked, either.

In 1855 the English astronomer royal George Biddell Airy (1801–1892) suggested that mountains and the rocks underlying them (their "roots") were lower in density than the rocks making up the lowlands. In fact, that was why, he decided, mountains were mountains. Wherever the rocky surface was lighter than the surrounding rocks, the surface floated higher. The lighter it was, the higher it floated.

The notion was carried further, in 1889, by the American geologist Clarence Edward Dutton (1841–1912). He felt that all rocks very slowly found their own level, depending on their density. He called the phenomenon *isostasy* and maintained that not only mountains but whole continents were made of lighter rock than the ocean basins were. That's why continents rose and *were* continents.

The continents, which were largely of granite, were thus floating on the denser basalt of the sea bottoms. And since they were floating, why might they not be (very, very slowly) drifting this way and that.

Given this new notion, an American geologist, Frank Bursley Taylor (1860–1938), reverted to the notion advanced by Snider-Pellegrini a half-century before. In 1898 he suggested that Africa and South America had split apart and were pulling away from each other while the relatively high ground in the center of the Atlantic Ocean remained put.

Taylor was definitely on the right track, but he, too, fell afoul of the problem of the mechanism. He suggested that the Earth had recently captured the Moon and that the sudden onslaught of huge tidal forces had split the supercontinent and forced its parts away from each other. Such a mechanism did not carry conviction, and it could not compete with the more popular land-bridge notion.

The idea was, however, taken further by the German scientist Alfred Lothar Wegener (1880–1930). He grew interested in the concept of isostasy and decided that this was really a death knell for the land-bridge theory. If there was a land bridge between Madagascar and India, it had to be composed of comparatively light rock. How could it then sink into the denser rock below? Even if something forced it down, it would surely bob up again. Wood doesn't sink and rise in the water; it always floats. And continents must always float, too. Therefore, if life forms had traveled between Madagascar and India, or between Africa and South America, it must have been because those land areas were at some time past, not thousands of miles apart, but in contact.

In 1912 he presented his "continental drift" suggestion as an alternative. He produced no mechanism, no Flood, no tidal forces. The continents just drifted. For evidence, he used the fit of the shorelines—and he fit them not at the actual shorelines but at the edge of the continental shelves and found the fit was even better then. He showed that polar areas had fossils of life forms that could not live under polar conditions and that seemed to make it plausible that the region had moved from a warmer latitude.

By 1922 he had succeeded in presenting evidence that at one time all the continents had clung together as a single huge land-mass, which he called *Pangaea* (Greek for "all land"). It was surrounded by a single huge ocean he called *Panthalassa* (Greek for "all sea").

Wegener also had a new explanation for mountain formation. By the old cooling-and-shrinking-Earth theory, mountains should have formed everywhere. If, however, one imagined the Americas drifting westward, then the leading edge meeting with some resistance from the ocean floor into which it drifted would wrinkle up into a mountain chain. That's why the Rocky Mountains and the Andes Mountains run parallel to the western shores of the Americas.

However, he had no mechanism that would serve to push the continents through the rock underlying the ocean basin, and everyone felt that that rock was too stiff for continents to push through, whatever the mechanism. The result was that despite all the favorable evidence that Wegener presented, he wasn't believed. Indeed, most geologists were rather ferociously against him and felt his theories were pseudoscientific nonsense.

Wegener was an enthusiastic explorer of Greenland. On a fourth and final trip there, he died on the ice cap in 1930. At the time of his death, his suggestion of continental drift was also virtually dead because of his lack of a reasonable mechanism for bringing about that drift.

When the answer did come, it came from the sea bottom.

In the 1850s, there was a huge attempt to lay a cable across the bottom of the Atlantic Ocean in order to allow direct telegraphic contact between the United States and Great Britain. For the purpose, information concerning the sea bottom was needed. An American oceanographer, Matthew Fontaine Maury (1806–1873), collected data on soundings of the ocean depths. In 1854, he noted that soundings in the middle of the ocean showed it to be shallower there than on either side. Apparently, there was a submerged plateau running down the center of the Atlantic, and Maury called it *Telegraph Plateau.*

There was, however, no chance of getting any fine detail concerning this mid-ocean plateau. The only way of determining depth, at that time, was to pay out several miles of weighted rope and measure the length after it hit bottom. It was a difficult technique, a lengthy and expensive one, and with the best will in the world, it would take years to do a few hundred soundings and that would not give much detail.

The turning point came during World War I, when Langevin developed sonar, as mentioned earlier. A beam of ultrasonic sound could be directed downward and would be reflected by the bottom of the ocean and returned. By measuring the time between emission and return the distance to the bottom could be calculated. Depth figures could be quickly obtained in any number, and a continuous profile of the sea bottom could be worked out.

The first oceanographic vessel to use this new technique was the German ship *Meteor,* which began its studies of the Atlantic Ocean in 1922. By 1925 it was clear that Telegraph Plateau was no mere plateau. It was a mountain range—longer, higher, and more rugged than mountain ranges on land. Its highest peaks broke through the water surface and appeared as islands—the Azores, Ascension, and Tristan da Cunha. The mountains were called the *Mid-Atlantic Ridge.*

After World War II, the task of continuing to study the ocean bottom fell chiefly to the American geologist William Maurice Ewing (1906–1974). By 1956 his sonar findings showed the Ridge was not confined to the Atlantic Ocean. At its southern end, it curves around Africa and moves up the western Indian Ocean to Arabia. In mid-Indian Ocean it branches, so that the range continues south of Australia and New Zealand and then works northward in a vast circle all around the Pacific Ocean. It was called the *Mid-Oceanic Ridge*, a 40,000-mile-long mountain range curving around the whole Earth.

What's more, the Mid-Oceanic Ridge was not like the mountain ranges on the continents. The continental highlands are of folded sedimentary rock, while the vast highlands of the ocean were of basalt squeezed up from the hot lower depths.

Ewing and his student, Bruce Charles Heezen (1924–1977), also discovered that along the center of the Ridge there was a deep canyon, and by 1957 it was clear that this ran the entire length of the Mid-Oceanic Ridge. It was called the *Great Global Rift*.

At first it seemed that the Rift might be continuous, a 40,000-mile crack in the Earth's crust. Closer examination, however, showed that it consists of short, straight sections that are set off from each other as though earthquake shocks had displaced one section from the next. And, indeed, it is along the Rift that many of the Earth's quakes and volcanoes tend to occur.

It appeared at once that the Earth's crust is divided into large plates, separated from each other by the Great Global Rift and its offshoots. These were called *tectonic plates* (from a Greek word for "carpenter," since the plates seemed to be cleverly joined to make a seemingly unbroken crust). The study of the evolution of the Earth's crust in terms of these plates is referred to by those words in reverse—plate tectonics.

What about Wegener's continental drift under these conditions? If an individual plate is considered, the objects upon it cannot drift or change position relative to that plate. North America is stuck forever on the plate that includes it (the North American Plate) in the position in which it is now. But what if the plate itself can move, carrying North America with it?

It might seem that this is unlikely since the neighboring plates are so tightly wedged together. Yet the plate boundaries are littered with volcanoes. Indeed, the shores of the Pacific, which make up the boundary of the Pacific Plate are so rich in volcanoes, both active and inactive, that the whole has been referred to as the "circle of fire."

Could it be, then, that heated fluid rock (magma) might force itself upward from the deeper layers of the Earth through the Rift in various places, making itself apparent in the form of volcanic action here and there? Specifically, magma might be welling up very slowly through the Mid-Atlantic portion of the Rift, appearing as active volcanic eruptions in Iceland (which lies on the Rift) but solidifying on contact with ocean water elsewhere. It could be that it was this magma that formed the Mid-Atlantic Range. Then, as more and more magma welled upward, the solidifying rock would force the North American plate and the Eurasian plate apart very slowly and similarly force apart the South American plate and the African plate as well.

It might have been, then, the upwelling through the Rift that broke up Pangaea and forced the portions apart, the separation steadily widening into the Arlantic Ocean. This is called *sea-floor spreading* and was first proposed by the American geologists Harry Hammond Hess (1906–1969) and Robert Sinclair Dietz (b. 1914) in 1960.

The continents are not floating or drifting apart, as Wegener had thought. They were fixed to plates that were themselves being pushed apart and were carrying the continents with them. This was a mechanism that could be demonstrated, and the world of geology, which had earlier scorned and derided Wegener, now flocked to the concept of Pangaea and its breakup with excitement and enthusiasm.

Naturally, if two plates are forced apart, each must (in view of the tightness of the fit of all the plates) be jammed into another on the other side. One plate must then slip under the other, dragging the sea floor down into *deeps*. Or else, two plates, jamming together, must crumple into mountain ranges.

About 225 million years ago, Pangaea began to break up into a northern half comprising North America, Europe, and Asia, and a southern half comprising South America, Africa, India, Australia, and Antarctica. The northern half is called *Laurasia* because the oldest part of the North

American continent is the Laurentian highlands north of the St. Lawrence River. The southern half is still called by Suess's name of *Gondwanaland,* but it includes no land bridges.

About 200 million years ago, North America began to be pushed away from Eurasia, and 150 million years ago, South America and Africa also began to be pushed apart. About 110 million years ago, the eastern portion of Gondwanaland broke into Madagascar, India, Antarctica, and Australia. Madagascar stayed fairly close to Africa, but India moved farther than any other land mass. It moved northward to push into southern Asia, forming the Himalayan Mountains, the Pamirs, and the Tibetan plateau— the youngest, greatest, and most impressive highland area on Earth. Antarctica and Australia may have separated only 40 million years ago, Antarctica moving southward to its frozen destiny.

The plates are still moving today, of course, and the continents are still slowly moving as a result. There is a great rift down eastern Africa, and the Red Sea may be the beginning of a slowly widening ocean. The continents may come together again hundreds of millions of years in the future, forming a new Pangaea, which may persist for some time before breaking up again to form new continents somewhat modified from the old. This may happen over and over in the same way the Pangaea of 225 million years ago may have formed from separate continents coming together—and there may have been another Pangaea long before that and still another long before that.

Plate tectonics has now proved to be the central core of the science of geology. It explains earthquakes, volcanoes, deeps, island chains, continental drift, the distribution of living organisms, and much more. It may even be that plate movements now and then push a continent across one of the poles, introducing glaciation and an ice age which may drop the sea level and cool the ocean waters, thus bringing on a mass extinction.

So we see that the Earth's surface evolves and that the

continents, as we now know them, slowly formed in the course of the Mesozoic and early Cenozoic eras.

And now, at last, we are ready to trace things further back and can ask about the beginnings of Earth itself.

# 14

## EARTH

Actually, there was no sensible way of estimating the age of the Earth until the uniformitarian principle was established. Once it was accepted that slow changes were taking place over long periods of time, the system for estimating Earth's age was clear. One must calculate the rate at which a particular slow change was taking place, determine the total change that has taken place, and then divide the latter by the former.

The first attempt to do this came in 1715, when the English astronomer Edmond Halley (1656–1742) reasoned as follows:

The rivers, as they flow, dissolve tiny quantities of salt from the land they flow through and deliver them to the ocean. The salt stays in the ocean, for only the watery portion of the sea evaporates under the action of the Sun. This water vapor falls as rain, which contains no salt to speak of, but as the rivers return the fallen water to the ocean, they deliver a bit more dissolved salt from the land. This happens over and over again.

If we suppose that the ocean was fresh water to begin

with, and if we measure how much salt is being added to it each year, then we can calculate how many years that salt must have been added to cause the ocean to be 3.3 percent salt, as it is today.

In principle, this is a straightforward and very simple arithmetical operation, but there were many gaping holes in it. First, it might be that the ocean didn't start as fresh water but had salt in it to begin with.

Second, it was quite impossible for Halley, in his time, to know the exact rate at which salt was being added to the ocean each year because many rivers outside Europe had never been chemically analyzed and even the volume delivered could not be accurately known. One had to estimate, judging by the rivers one knew, and the estimate might easily be wildly wrong.

Third, there was no way of knowing whether the rate of salt delivery to the ocean really remained constant year after year. Rivers might be more turbulent or more placid at certain periods of Earth's lifetime, and the present state might be nowhere near the average.

Fourth, there were processes that could remove salt from the ocean. Storm winds send ocean spray, with its salt content, over the land. Shallow arms of the ocean can dry up completely, leaving their salt content behind (which is where salt mines come from). Taking this all into account, it would therefore be quite possible for Halley to end up with a figure that was dreadfully wrong.

His estimate was, in fact, that Earth's ocean might be as much as 1,000 million years old. This, actually, was quite a respectable estimate for the first time round. At the time, though, it made little imprssion. Ussher's decision still held sway at that time, and it was easy to maintain that when God created the Earth 6,000 years ago, he created it with an ocean containing today's level of salt.

(Indeed, every once in a while people argue in this way against the evidence presented in favor of biological evolution. God created the Earth, they say, with all the fossils already in place and with all the other evidence of a long age for the Earth as well. This was done either to fool humanity, out of a malicious sense of humor, or to test people's faith in revelation over observation and reason, or for other trivial un-Godlike motivations. Some who are wedded to the literal words of the opening portion of the Bible might accept this sort of argument, but thinking people, even if sincerely religious, do not.)

Another way of estimating the Earth's age depended upon sedimentation rates. The rivers, lakes, and oceans of the world laid down mud and sludge—*sediment*—and such sediment, under the weight of further layers laid down about it, was compressed into *sedimentary rock*. Since the watery parts of the globe were rich in life, it frequently happened that living things, recently dead ones, or parts of them were trapped in the sediment under conditions that made for fossilization. Even land animals had to find water periodically and might be trapped in waterholes, or killed there, and somehow end up in the sedimentary rock as fossils.

Fossil hunters could measure the thickness of the sedimentary rock in which they found fossils. If the rate of sedimentation could be determined, then from the thickness of the strata representing a particular geological period the duration of that period could be calculated. Once the periods were put in order, the total duration for all of them and the time lapse since the present could be determined.

This was not a very accurate way of measuring the age of the fossils, for it was impossible to say whether the sedimentation rate was the same in one place as in another, or at one time as in another. Variations were so great (and sometimes not really known) that no calculated average could really be trusted.

Still, estimates were advanced to the effect that the oldest fossils were perhaps 500 million years old, and that was not at all bad for dealing with something as uncertain as sedimentation. It was against this background of an Earth that was possibly 500 million years old or more that Darwin was able to postulate a scheme of biological evolution involving random variations, with natural selection serving to remove the randomness and to lend the process the illusion of purpose. This was bound to be a very slow process and it needed hundreds of millions of years of time.

Yet even before Darwin presented his theory, this notion of an Earth that was extremely old was contradicted not out of religious considerations, but by scientists making use of apparently incontrovertible physical laws.

In the 1840s it was becoming more and more clear that energy could neither be created nor destroyed. The Universe, it seemed, had a fixed supply of energy, which could be converted from one form to another, but which remained unchanged in total amount. This is called the *law of conservation of energy* or *the first law of thermodynamics* and is, to this day, considered the most basic of all the laws of nature. It was formally stated by the German physicist Hermann L. F. von Helmholtz (1821–1894) in 1847.

Once the law of conservation of energy was presented and accepted, there arose the question as to the source of the Sun's energy. The question had never arisen before. It was thought that either the Sun shone steadily, day after day, through all of history because that was the will of God, or it was simply a ball of light that, by its very nature, glowed forever.

That couldn't be. If the Sun were a natural phenomenon, then it had to be emitting vast quantities of energy to light and warm the Earth from a distance of 93 million miles (150 million kilometers), and that energy had to come from somewhere.

The Sun could not get its energy as earthly fires did. The fires on Earth arose from the chemical combination of fuel and oxygen. If, however, the Sun consisted of fuel and oxygen, then all its content, even though it is 333,000 times the mass of the Earth, would be burned up in less than a third of historic times if it kept producing energy at its present rate.

Some other, and greater, source of energy had to be responsible for the Sun. By 1854 Helmholtz had decided that only one source of energy was great enough, and produced little enough change in the Sun, to account for its energy production. The Sun, he decided, had to be contracting. Its substance was falling inward and this fall represented a loss of gravitational energy that was converted into radiation that reached Earth as light and heat.

A contraction of 1/2000 of the Sun's radius would have supplied all the energy it had poured out since the Sumerians invented writing. Such a contraction would have gone unnoticed to the unaided eye and so everything seemed well.

This means that when the Sumerians invented writing 5,000 years ago, the Sun was just a trifle larger in reality and, therefore, in appearance than it is today, and if one went back an additional 5,000 years to the beginning of civilization, it would be another trifle larger and so on.

The Scottish physicist William Thomson, Lord Kelvin (1824–1907) took up the matter and by 1862 had calculated that 50 million years ago the Sun had extended out to Earth's orbit. In other words, if the Sun had started out filling Earth's orbit and had contracted to its present size it would have emitted energy at its present rate for only 50 million years. That meant that earth had to be no more than 50 million years old and could not have supported life until the Sun had contracted sufficiently to leave Earth comparatively cool. Life, then, would be far less than 50 million years old.

This horrified both geologists and biologists, who were absolutely certain that the Earth was far older than that. Kelvin's suggested age was, by his time, as ridiculously small to those who studied the slow changes in the Earth's crust and in evolutionary development as Ussher's suggested age was.

Yet how could one argue with the law of conservation of energy? All that the biologists and geologists could do was to insist that somewhere, somehow, there was another source of energy, one that was bigger and better than solar contraction, that would account for the Sun's energy over at least ten to twenty times the period that Kelvin allowed.

The solution both to the age of the Earth and to the energy source of the Sun arose out of a discovery by the French physicist Antoine Henri Becquerel (1852–1908).

In 1896 he accidentally discovered that the element uranium slowly but steadily gave off energetic radiations. The Polish-French physicist Marie Sklodowska Curie (1867–1934) discovered in 1898 that the element thorium also gave off energetic radiation, and she named the phenomenon *radioactivity*.

The uranium and thorium (as well as other elements and element varieties found to be radioactive), in giving off this radiation, were producing energy. Pierre Curie (1859–1906), the husband of Marie, was the first, in 1901, to measure the energy production, and he was able to show that the total energy a given weight of uranium emitted was enormously higher than the energy given off by the same weight of burning coal. The radioactive energies are given off so slowly, however (over a period of thousands of millions of years in the case of uranium and thorium), that only delicate measurements reveal its existence.

The New Zealand–born British physicist Ernest Rutherford (1871–1937) suggested in 1904 that this new source of energy, in some form, must be the answer to the problem of the Sun's energy. It was so incredibly rich a source that it would allow the Sun to shine for billions of years without perceptible change. That would allow the Earth to be as old as geologists and biologists said it was. He said this in a public lecture with the aged Kelvin himself in the audience.

But what was the precise source of this energy of radioactivity? None was apparent at first. Did this mean, then, that the law of conservation of energy would have to be abandoned?

It did not have to be. Rutherford allowed radioactive radiations to smash into intact atoms and the result made it plain that the atom was not just an ultra-tiny featureless ball as chemists had assumed it to be throughout the nineteenth century. By 1911, he showed that atoms consisted of a *very* tiny nucleus at the center, a nucleus only

1/100,000 the diameter of the atom as a whole. Almost all the atomic mass is in that tiny nucleus. Around it, filling the rest of the atom, is a froth of light electrons.

Ordinary energy obtained from chemical change, such as in the burning of fuel or in the explosion of dynamite, results from alterations in the arrangements of the light electrons. The much greater energies of radioactivity result from alterations in the much more massive particles within the tiny nucleus. In this way, *nuclear energy* was discovered.

Clearly, then, the Sun must be powered by nuclear energy, though the exact details were not worked out for another twenty years.

And as though that were not enough, the phenomenon of radioactivity served another purpose, too, that was in its way just as exciting.

Scientists quickly discovered that when a radioactive atom gave off energetic radiation, its nucleus rearranged itself so that the atom became different in nature. In 1904, the American physicist Bertram Borden Boltwood (1870–1927) pointed out that as uranium (or thorium) broke down it formed another kind of atom that also broke down, giving off radiations to form a third kind that broke down, and so on. Thus, one could speak of a *radioactive series*. Boltwood also pointed out that the final atom in both the uranium series and the thorium series was lead. The lead atom that was produced in the series was not radioactive and it changed no further. The net effect of this kind of radioactivity, then, was to change uranium or thorium into lead.

In that same year of 1904, Rutherford showed that a particular radioactive substance always acted so that half of any quantity always broke down in the same particular length of time. This length of time he called the *half-life*. (The concept has already been mentioned earlier in the book in connection with carbon-14.)

Each different radioactive substance has a different half-life, in some cases a tiny fraction of a second; in others, thousands of millions of years; and in still others, anywhere in between. A given substance always has the same half-life, at least under earthly conditions. If the half-life of a particular radioactive substance is known, it is easy to calculate how much of it will be left after any given time.

Boltwood suggested in 1907 that if a rock contained uranium, some of it was bound to be very slowly transforming itself into lead. From the amount of lead that had accumulated in the rock in association with the uranium, you could calculate how long the rock had existed in solid form. (As long as the rock was solid, neither the uranium nor the lead could escape from it.)

Since the half-life of uranium is 4,500 million years and that of thorium 14,000 million years, then even if the Earth were many thousands of millions of years old, not all of the uranium or thorium would have had time to break down and you would still be able to calculate the age of the rock.

As it happens, uranium and thorium are present in a wide variety of Earth's rocks, so that almost any of them can be easily dated. To be sure, the uranium and thorium are present in small quantities, but the detection of radioactive substances is a very precise procedure and small quantities are all that are needed.

As time went on, other radioactive substances were discovered with half-lives in the thousands of millions of years. The very common element potassium has a particular variety, potassium-40, that makes up about one out of every 10,000 potassium atoms. Potassium-40 is radioactive and has a half-life of 1,300 million years, breaking down to argon-40, a gaseous substance that is stable.

Another element, rubidium, that is less common than potassium, has fully one-quarter of its atoms as rubidium-87, which is radioactive and has a half-life of 46,000 million years, breaking down into strontium-87, which is stable. Both potassium and rubidium can also be used to determine great ages with considerable accuracy.

(Incidentally, the fact that such radioactive substances are widespread in the Earth's crust is of importance. They are not present in sufficient quantity to be terribly harmful to life. After all, life has lived with these radioactive substances for a long time and has not been wiped out. However, these radioactive substances act as a low, but very long-lived source of heat that accumulates in the Earth's crust perhaps as fast as the Earth radiates heat to space. This means that the Earth is cooling down only very slowly, if at all, and completely wipes out any geologic theories that involve the kind of cooling and shrinking of the Earth that would take place if there were no longtime source of heat within the planet.)

To be sure, although the principle of age measurement by radioactive breakdown is quite simple and straightforward, the practice can be difficult. Rocks have to be sampled carefully, delicate radioactive measurements have to be made over and over, there has to be some way of determining whether any lead (or strontium, or argon) was present to begin with, having no relationship to radioactive breakdown, and so on.

Nevertheless, methods were worked out and made practical, and the durations of the various geological periods, and the time before the present in which they existed, were calculated. The figures given in previous chapters were obtained in this way.

In fact, rocks were discovered that were older than any that we have considered so far. There were rocks that were 1,000 million years old, and by 1931 rocks had been found that were 2,000 million years old—and older, too. Particularly old rocks in western Greenland topped the 3,000 million year mark. The oldest rock so far found seems to be 3,800 million years old, give or take a hundred million years.

This represents a *minimum* age for the Earth, for the older a rock is the less likely it is to be found reasonably untouched during all its existence. Rocks may be eroded by the action of wind, water, or life; or they may be carried far down into the earth by plate movement and melted. It may be, then, that rocks older than 3,800 million years exist but are so rare that they have not been found, or perhaps, indeed, no rocks have survived for longer than that period.

Nevertheless, scientists have been able, from the changing proportions of rubidium and strontium in rocks, to reason out when the Earth first assumed something like its present size and structure. What seems most likely now is that the Earth formed 4,550 million years ago.

Such a figure gives us a completely different perspective on geologic time. When I said, in a previous chapter, that the first chordates appeared 550 million years ago, that would naturally seem like an event that had taken place in an unimaginably distant past. And yet, as a matter of fact, we now see that it happened rather recently. Moving 550 million years into the past take us only through the last eighth of Earth's history. For the first seven-eighths of its existence, there were no chordates of any kind, not even the simplest, living anywhere.

But, as I said, even at the time the chordates appeared, life was flourishing in the Cambrian seas. All the other phyla were in existence. There must be fossils, then, of organisms that are much older than any chordate and there is certainly plenty of room in Earth's history for them to have appeared. Let us not, then, ask for the beginning of one particular phyla or another, but for any of them.

What was the beginning of fossils, generally?

# 15

## FOSSILS

The common fossils in the Cambrian period, which was between 600 million and 500 million years ago, are *trilobites*, so called because their bodies consist of three lobes and are, therefore, trilobar. They are arthropods, the phylum to which modern crustaceans such as crabs and lobsters belong, and to which such land organisms as insects and spiders also belong.

About 10,000 species of trilobites have been found, some as small as a tenth of an inch (2.5 millimeters) long and some over two feet (26 centimeters) long. They suffered terrible losses in a couple of mass extinctions during the Cambrian from which they were finally unable to recover. After the Cambrian, they rapidly diminished in numbers and all were gone before the end of the Paleozoic era.

They did leave an echo behind, though. There is the horseshoe crab, which has now existed with little change for as much as 200 million years, since the Jurassic. They are to trilobites as crocodiles are to dinosaurs. (Structurally speaking, the horseshoe crabs, and the trilobites too, are more closely related to spiders than to crabs.)

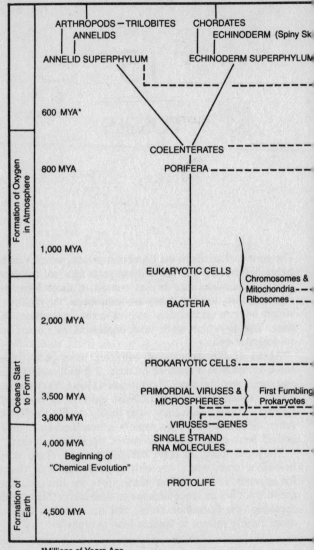

# PRIMITIVE OCEAN LIFE
(Pre-Cambrian)

ARTHROPODS — TRILOBITES      CHORDATES
ANNELIDS      ECHINODERM (Spiny Sk
ANNELID SUPERPHYLUM      ECHINODERM SUPERPHYLUM

600 MYA*

Formation of Oxygen in Atmosphere

COELENTERATES
800 MYA      PORIFERA

1,000 MYA

EUKARYOTIC CELLS      Chromosomes & Mitochondria
BACTERIA      Ribosomes

2,000 MYA

PROKARYOTIC CELLS

Oceans Start to Form

3,500 MYA      PRIMORDIAL VIRUSES & MICROSPHERES      First Fumbling Prokaryotes
3,800 MYA
VIRUSES — GENES
4,000 MYA      SINGLE STRAND RNA MOLECULES
Beginning of "Chemical Evolution"

PROTOLIFE

Formation of Earth

4,500 MYA

*Millions of Years Ago

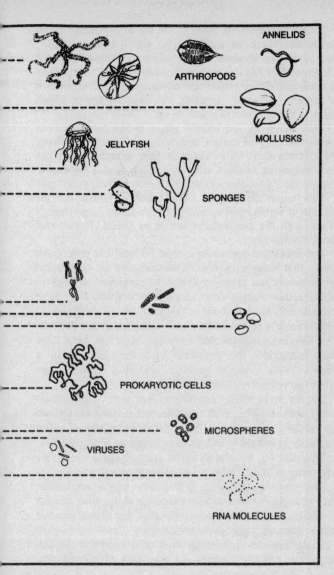

ANNELIDS

ARTHROPODS

MOLLUSKS

JELLYFISH

SPONGES

PROKARYOTIC CELLS

MICROSPHERES

VIRUSES

RNA MOLECULES

Fossils representing other phyla are also present in the Cambrian. There are mollusks (of which oysters, clams, and squid are modern representatives), echinoderms (of which starfish and sea urchins flourish today), brachiopods (a kind of shellfish rather rare nowadays), poriferans (such as modern sponges), annelids (the best known of which today is the earthworm), and so on.

Very likely all the animal phyla except the chordates were present in the early Cambrian, and, of course, simple plant forms also. All of them, in fact, stretch back to the very beginning of the Cambrian (which is also the beginning of the Paleozoic), about 570 to 600 million years ago.

Now come the puzzles. The Cambrian rocks are the earliest in which copious fossils of life forms large enough to see with the unaided eye are to be found. Before that there is nothing.

The rocks that are younger than 600 million years have fossils that because of mass extinctions and the consequent proliferation and rapid evolution of survivors, change in nature rather sharply from stratum to stratum. It is these more or less sudden changes that first caused geologists to divide Earth's recent history into periods and sub-periods. The Paleozoic is separated from the Mesozoic by a huge mass extinction, the Mesozoic from the Cenozoic by a mass extinction almost as huge, and finer divisions are often marked off by lesser extinctions.

But for rocks older than 600 million years, there are no fossil markers. The older rocks are not divided neatly into periods and sub-periods. The most frequent way of referring to these old rocks and strata is simply as *pre-Cambrian*.

Why did all this fossilization arise so suddenly at the beginning of the Cambrian out of (apparently) nothing?

One possible suggestion is that some supernatural influence brought life into sudden existence at *this* time and not in 4004 B.C. and that it was only after this divine creation that evolutionary processes took over.

That, however, is a suggestion of desperation. In science, one always assumes the operation of natural processes. For instance, we know that the fossils we find are chiefly of the hard parts of organisms—teeth, claws, bones, shells and so on. For this reason, it is quite possible that fossils don't give a completely true picture of the relative importance of life forms in different eras. Those phyla that have bones (chordates) or shells (arthropods, mollusks, and so on) may well be overrepresented. Those phyla or portions of phyla in which hard parts are rare or totally absent leave traces that are rarely encountered in the fossil record, and they are harder to recognize when found.

It may be, then, that hard parts were only evolved at the beginning of the Cambrian and that it is then that fossilization began to leave its mark. This sounds like a reasonable thought, but it leaves us with the problem of explaining just why hard parts developed so suddenly at this particular time. (We'll try our hand at this later in the book.)

It should also be remembered that all the phyla seem to have been well evolved by Cambrian times. They seem distinctly separate, even the chordates who first developed some way into the Cambrian.

If we dismiss any possibility of a supernatural intervention that created life *already* separate, then, by evolutionary principles, we can only suppose that there was a long development before the Cambrian period, during which the separate phyla split off from some ancestral stock. We can't follow the details of such development because of the lack of pre-Cambrian fossils, but surely we can reasonably suppose that the development took place.

This notion of pre-Cambrian development began to seem all the more likely when the true age of the Earth was finally determined. Since the Earth is 4,550 million years old, the pre-Cambrian period is just about 4,000 million years in duration and makes up seven-eighths of the entire history of the Earth. Clearly, there was plenty of time for the slow development of the different phyla.

In order to investigate this possibility, let's take up next the beginnings of multicellular organisms.

# 16

## MULTICELLULAR ORGANISMS

As I mentioned earlier, the notion of the evolutionary development of life arose, to a great extent, from the observation of similarities between animals. Wolves and jackals are similar, as are sheep and goats, lions and tigers, horses and donkeys, and so on. Going further, groups of groups are similar in some more fundamental ways, and groups of these larger groups are similar in still more fundamental ways, and so on. The most logical way of explaining this (short of supposing some supernatural agency who created life in this fashion in order to mislead us) is to presume evolutionary development and to weigh the evidence with that in mind.

If, however, there are similarities which are quite patent to the unaided eye, there should be further similarities, and extremely fundamental ones, perhaps, that might become obvious if we could see tiny details that the unaided eye cannot make out.

There are ways of magnifying the appearance of things. Spheres of glass seem to enlarge the appearance of what they rest on, as do drops of water. Such magnification is small and poorly focused, however. What was needed was some device, deliberately created by human beings, that would magnify clearly, and to a large extent.

The first hint of this came after Galileo constructed a telescope in 1609. It magnified things at a distance and enabled him to study astronomical objects in greater detail than had hitherto been possible. He found that by a proper rearrangement of lenses, he could also magnify the appearance of small objects. Thus, he had what came to be called a *microscope* (from Greek words meaning "to see the small") and used it to study insects.

This was just a passing observation of Galileo. The first to take up microscopy in all seriousness was an Italian biologist, Marcello Malpighi (1628–1694). Beginning in the 1650s, he used a microscope to investigate the lungs of frogs and the wing membranes of bats. From such observations, he discovered tiny blood vessels *(capillaries,* from Latin words meaning "hairlike") that were invisible to the unaided eye and that connected arteries and veins. He also studied insects and chick embryos, and others quickly followed his example.

In 1665 the English scientist Robert Hooke (1635–1703) studied a thin slice of cork under a microscope and found it to be composed of a finely serried pattern of tiny rectangular holes. These, Hooke called *cells,* a term till then generally used to signify small rooms.

Cork is a dead tissue, however. Living plant tissue is also composed of those small units, but these units are filled with a complex fluid. The name of cells still applies to them, however, though, strictly speaking, the term is now a misnomer.

Cells in living tissue were observed now and then, but it was not till 1838 that a German botanist, Matthias Jakob Schleiden (1804–1881), stated that as a general rule all plants consist of cells.

Plant cells are separated by pronounced *cell walls* containing cellulose, a supporting substance characteristic of all plants but not found in animals. Animals also possess cells but these are separated from each other by relatively thin *cell membranes*. In 1839, the German physiologist Theodor Schwann (1810–1882) maintained that all animals were made up of cells.

Between them, Schleiden and Schwann established the *cell theory* of life.

All the animals I've mentioned so far are *multicellular*. That is, they all consist of a number of cells. Often, it is a very large number. A large whale might be made up of a hundred quadrillion (100,000,000,000,000,000) cells; a human being of fifty trillion (50,000,000,000,000). Even the tiniest insect, however, though it may be made up of only a few thousand cells, is still a multicellular animal. The plants we see growing on land are multicellular, too.

Plant cells and animal cells are easily distinguished. Plant cells have cell walls and animal cells have cell membranes. In addition, many plant cells have chlorophyll, which is contained in small structures called *chloroplasts* (Greek for "green forms"), while animal cells never have chlorophyll.

Nevertheless, plant cells among themselves, and animal cells among themselves, are surprisingly similar. To be sure, in a single organism such as a man, muscle cells are quite different in appearance from nerve cells, and both differ from liver cells. However, nerve cells from one kind of animal are quite similar to nerve cells from another kind of animal, and the same is true of other particular types of cells. Even when organisms are quite different in appearance and belong to different phyla, the cells are similar in size, appearance, and structure; certainly much more similar than the organisms themselves are.

The similarity in cells throughout all the phyla is strong evidence in itself that the phyla have a common ancestry. If the phyla came into independent existence through distinctly different evolutionary processes, we might expect that some phyla might not be made up of cells but have some different organization; or that if two phyla were both made up of cells, the two would have cells of radically different size or appearance. This, however, is not so, and if we consider the chemical makeup of all cells (which we will have occasion to do later in the book) we would see that the similarities among them are even closer.

It becomes reasonable to suppose then, that *all* forms of life, however different overall in size, appearance, structure, and function, are descended from some common ancestor. We can't very easily study the rocks for traces of the details of that descent (though the case is not totally hopeless, as we shall see), but we can at least study the organisms that now exist for any clues as to the nature of the descent.

Thus, all multicellular organisms can start life as a single cell. There are seeming exceptions of course. A plant can start from a twig, which is already multicellular. A portion of a starfish, already multicellular, can give rise to a complete one. This sort of reproduction is called cloning.

In general, cloning is only found among plants and the less complex animals. Complex animals, in nature, start only from an egg, and even those plants and simple animals that tend to clone can also start from an egg, a seed, or a spore.

In 1861 the Swiss physiologist Rudolf Albert von Kölliker (1817–1905) clearly showed that mammalian eggs and sperm had the structures that were characteristic of single cells. We therefore speak of an *egg cell* or a *sperm cell*. The union of an egg cell and a sperm cell forms a *fertilized ovum* (*ovum* is Latin for "egg") and that, too, has the structure of a single cell. It is from the fertilized ovum that an organism as small as a shrew or as large as a whale develops.

The German anatomist Karl Gegenbaur (1826–1903), a student of Kölliker, went on to show that all eggs and sperm, even the giant eggs of reptiles and birds, were single cells. The egg of a bird or a reptile contains a tiny speck of life that is the fertilized ovum itself, and all the rest is a food supply for the developing embryo.

The fertilized ova of different animals are often quite similar in appearance. It is almost impossible to tell the fertilized ovum of a giraffe from that of a human being by ordinary microscopic appearance alone. There is a difference, of course, for one produces a giraffe and one a human being, without possibility of a mistake, but the difference exists at the molecular level and is quite sub-microscopic.

Cells have the ability to divide into two as a result of complex processes involving the structures within the cells (the details of which we needn't go into now). These processes are essentially the same in all cells, another strong piece of evidence for the descent of all life from a common ancestor.

The fertilized ovum divides into two cells, which divide into four, which divide into eight, and so on. In the process, individual cells gradually specialize and become the ancestors of particular tissues and organs in the final animal that forms. The details of such development can give a notion of relationships.

For instance, some animals, in the course of their development, appear in a youthful form that is somewhat (and sometimes very) different from the adult form. The best-known case is the caterpillar, which, after eating and growing, forms a cocoon within which its body is reorganized so that it is born again as a butterfly. The youthful form, when so different from the adult, is called a *larva*. (*Larva* is a Latin word of which one meaning is "mask," since the larval form effectively masks the adult form into which it eventually turns.)

Among land vertebrates, larval forms are not found, but the tadpole is a well-known larval form of a frog or toad.

Certain organisms that in adult form are *sessile,* or fixed in place (like oysters, for instance), have larval forms that swim freely about and select spots (as far as we can speak of "selection" for an organism as unintellectual as that of an oyster larva) on which they can settle down to adult immobility.

In general, adult forms are more specialized than larval forms are, and it is therefore the larval forms that are liable to give some hints as to the ancestry of a particular organism. Thus, from oyster larvae we can reasonably suspect that oysters are descended from free-swimming ancestors.

Starfish have *radial symmetry*. That is, from the center of the organism there are repeated parts radiating in all directions. In the case of ordinary starfish there are five rays pointing outward at equidistant intervals (and in some species more than five). Starfish are *echinoderms* (Greek for "spiny skins"), and there are echinoderms called sea urchins that don't have the obvious radial symmetry that starfish do, but that turn out to have it on closer examination.

Radial symmetry is a rather primitive property. All the phyla but the very simplest have bilateral symmetry, in which the body can be divided (in imagination) into two sections, lengthwise, with the left section the mirror image of the right section. We (and all vertebrates) are bilateral in this sense, so that an organ on one side is matched by an organ on the other. We have two eyes, two shoulders, two breasts, two nostrils, two lungs, two kidneys, and so on. Any organ we have one of is more or less along the central line of the body: one nose, one heart, one navel, one larynx, and so on.

But are starfish really very primitive because of their radial symmetry? No, for the radial symmetry is a specialization that developed late in their evolution. We know this because the larvae of echinoderms are as bilaterally symmetric as we are. The echinoderms developed from a bilateral ancestor.

Can we tell much from chordate larvae? We might think not, since larvae are not common in our phylum. Even the tadpole is a late development and only tells us that amphibia are descended from fish.

The simple chordates, however—those that are not vertebrates—have larval forms that may be significant. The tunicates, for instance, are as immobile as oysters and were originally considered to be mollusks when they were first discovered, before the significance of their gill slits was grasped. The tunicate larvae, however, are free-swimming and look rather like the amphioxus.

(One way in which evolution might take place is by a

phenomenon called *neoteny*—from Greek words meaning "stretched-out youth"—in which the larval stage becomes more and more important. Perhaps some early tunicates developed larval forms that never did turn into adults but developed sexual organs so that from them developed aphioxus-like organisms and from these the vertebrates. That, however, is just speculation.)

The most interesting larval form is that of balanoglossus, which may well be the most primitive of all chordates living today. The balanoglossus larva is so similar to echinoderm larvae that the former was classified with the echinoderms before the adult form was identified.

The similarity of the larval forms of balanoglossus and echinoderms makes it seem possible that some ancestral form evolved in two directions. In one, it gradually grew more and more echinodermish, developing radial symmetry. In the other, it grew more and more chordatish, eventually developing bone.

Starfish and humans, however, are so different that it seems rather difficult to suppose that there is a common ancestor. That is a lot to accept on the basis of larval forms involving the balanoglossus, which seems, at best, to be no more than half-chordate. (In fact, the sub-phylum that includes balanoglossus is called just that, *Hemichordata.*)

Is there anything else? We might try some chemical characteristics that would differentiate chordates from other phyla.

For instance, there is an important compound in our muscles that is intimately connected with the mechanism whereby muscles contract and relax. It is called *creatine phosphate,* and it can be abbreviated as CP. CP is found in all vertebrate muscles, without exception. Muscles in other phyla, however, do not have CP; instead they have a somewhat similar compound called *arginine phosphate* or AP.

What about those chordates that are not vertebrates. Amphioxus has CP; tunicates have AP; balanoglossus has CP *and* AP.

And the echinoderms? Most of these have AP only, but

sea urchins have CP *and* AP, and brittle stars (which resemble starfish except that the arms are longer and more flexible and emerge from a globular little body) contain CP.

It might seem, then, that somewhere along the line the common ancestor of the echinoderms and chordates, after it had begun to evolve slightly divergent species, developed the use of CP. This use survived in a few of the echinoderm species as they evolved and in all the chordates above the level of the tunicates.

The echinoderms and the chordates, then, together make up the *echinoderm superphylum*. The superphylum (a division containing more than one phylum) is named for the echinoderms because that is the more primitive of the two and the common ancestor must have been more echinodermlike than chordatelike.

The echinoderms and the chordates differ in that the chordates are *segmented* and the echinoderms are not. By segmented, we mean that an organism is made up of a number of connected and similar parts, each multicellular. These parts are called segments, and certain organs are repeated in each.

In chordates such as ourselves, the segmentation is not immediately obvious, but if we look at a human skeleton, the backbone and the ribs are clearly examples of segmentation. The arrangement of muscles and nerves shows segmentation, too, as do other types of organs, in the course of embryonic formation if not in the adult.

There are two other segmented phyla, the annelids and the arthropods. Neither one shows any close relationship to the chordates in any other respect, so that it is customary to assume that the trick of segmentation was evolved at least twice, once by the chordates and once by some common ancestor of the annelids and the arthropods, if *they* are related.

Biologists judge that the annelids and arthropods *are* related because of a number of basic similarities and because there are a numer of species of animals called *peripatus*, which have both annelid and arthropod characteristics. The peripatus seems to be a descendant of a common ancestor of the annelids and arthropods and has kept many of the primitive traits of that ancestor; just as the balanoglossus may be a descendant of a common ancestor of the echinoderms and chordates.

Therefore the annelids and the arthropods make up the *annelid superphylum*. It is named for the annelid because that is the more primitive phylum of the two, and the common ancestor closer to the annelid than to the arthropod in characteristics.

As the fertilized ovum grows and develops in multicellular animals, it eventually forms a ball of cells with a space in the middle. A portion of the ball then caves in to form a cup-shaped object, with two layers of cells, one facing the outside world and one facing the inside of the cup. The one on the outside is the *ectoderm* (Greek for "outer skin") and the one on the inside is the *endoderm* ("inner skin").

The ectoderm and endoderm are called *germ layers*, from an old meaning of "germ" as a small bit of life. As the organisms continue to grow and differentiate, such organs as the skin, nervous system, and sense organs form from the ectoderm. From the endoderm, there develop such organs as the stomach and intestines, the lungs, and the digestive glands.

In all but the very simplest phyla, a third germ layer develops between the ectoderm and endoderm. This is the *mesoderm* or middle skin and from it develop muscle, blood, connective tissue, and kidneys. And that is all; no phylum has ever developed a fourth germ layer.

The mesoderm is formed in one of two ways. It can form from pouches growing out of the endoderm. Or it can form from the place at which the endoderm and ectoderm meet. Only in echinoderms and chordates (in the echinoderm superphylum, in other words) does the mesoderm arise from the endoderm only. This is another piece of evidence of the relationship between the echinoderms and the chordates.

In all the other phyla that have mesoderms, the mesoderm develops from the ectoderm-endoderm junction. For that reason, all the other mesoderm phyla are included in the annelid superphylum.

Thus, all the phyla with three germ layers seem to have arisen from one of two ancestral forms, each of which evolved independently a different way of forming the mesoderm. From one, there arose the echinoderms and chordates and from the other all the rest. (It seems to me that an extraterrestrial intelligence surveying Earthly life would conclude that the annelid superphylum, from the number of its phyla and species, was by far the more successful of the two. Naturally, we, from our position in the smaller of the two superphyla, find it difficult to agree with that.)

But from where did these two ancestral forms of the superphyla arise? There does exist, even today, a primitive phylum that makes do with only two germ layers, an ectoderm and an endoderm—the *coelenterates* (Greek for "hollow gut"). They are essentially cup-shaped groups of cells, rather resembling the cup that is formed in the course of the development of the more advanced phyla—the cup that precedes formation of the mesoderm.

The coelenterates have a single opening to the cup that serves as both mouth and anus. Food is taken into the inside of the cup (the "hollow gut") through the one opening. There it is digested, and the wastes are then ejected through that same opening.

The best-known coelenterates of today are such animals as jellyfish, coral, and sea anemones. These must be descended from very primitive coelenterates that were once the most complex animals in existence. Some of the early descendants, however, branched off to develop a mesoderm in each of two different ways, thus giving rise to the two superphyla that overwhelmingly outrange in importance those few organisms that have continued to cling to the coelenterate way of life.

Even more primitive than the coelentarates are the *porifera* (Greek for "pore possessing") or sponges, which are barely multicellular. Sponges consist of a sessile structure full of pores. Water is sucked through the pores, and from it edible bits of life are digested and the remains ejected through certain larger holes.

Although there are several specialized types of cells in sponges, the specialization has not gone very far. In truly multicellular animals, individual cells are so specialized as to be dependent on their neighboring cells to perform other functions also necessary to themselves. The result is that individual cells of a multicellular organism cannot live and grow on their own, but die if separated from the organism. In sponges, on the other hand, each individual cell can, on its own, multiply and give rise to a new sponge.

There are other cases of such limited association short of true multicellularity. The various seaweeds are examples of limited association of plant cells.

The question now is, if the phyla of the Cambrian originated from the ancestors of the two superphyla, and if these ancestors were descended from primitive coelenterates, the earliest *true* multicellular organisms, when was this beginning of multicellularity?

Some pre-Cambrian traces of multicellular life have actually been detected. In 1930 a German paleontologist, Georg Julius Ernst Gurich (1859–1938), found indisputable traces of multicellular life in rocks that just antedated the Cambrian. In 1947 an Australian paleontologist, R. C. Sprigg, found traces, not of material fossils themselves, but of impressions on late pre-Cambrian rocks that were left by soft-bodied multicellular animals. These were identified as including worms, jellyfish, and sponges, the most primitive of all the multicellulars.

We can't get enough detail to be able to spot a beginning directly. However, paleontologists have come to certain conclusions as to the rate of evolutionary change. Using those conclusions, they suspect that the first multicellular organisms came into being about 800 million years ago. These simple organisms, consisting of soft parts only, persisted for about 200 million years (one fourth of the total existence of multicellular organisms) before hard parts were developed and true fossilization began.

Yet multicellular organisms did not spring out of nothing, either. Before they existed there must have been still simpler organisms made up of single cells of the kind that, in the course of evolution, eventually came together to form multicellular organisms. Such cells are called *eukaryotic cells* for reasons I will soon explain. An organism made up of a single eukaryotic cell is a *eukaryote*.

Consequently, we must now turn in that direction and probe the beginnings of eukaryotes.

# 17

## EUKARYOTES

When the cell was first recognized it seemed to be a microscopic body filled with a gelatinous fluid in which few or no details could be made out by the microscopes of the time.

In 1839, just about the time the cell theory was being advanced, a Czech physiologist, Jan Evangelista Purkinje (1787–1869), used the term *protoplasm* for the bits of life in eggs. The word is Greek for "first-formed," since the embryonic material is an individual creature's first form of life, something that eventually grows, divides, and differentiates into a complete adult organism.

The German botanist Hugo von Mohl (1805–1872) made use in 1846 of the same term (perhaps without knowing of Purkinje's earlier use) for the gelatinous material inside any cell. By 1860 the German anatomist Max J. S. Schulze (1825–1874) had demonstrated that protoplasm has similar properties in all cells, whether of complex organisms or of very simple ones, and whether plant or animal. This helped show that all life on Earth is essentially one and made the case for evolution stronger.

And yet one could not imagine protoplasm to be a uniform jelly that was *entirely* the same in all cells, since, in that case, what would it be that made it possible for each organism to give birth to young of the same species as itself? There had to be something in the protoplasm that unfailingly distinguished each species.

Actually, the existence of one structure within the cell was discovered even before the word *protoplasm* was invented. In 1831 a Scottish botanist, Robert Brown, was studying the cells in orchid leaves and discovered that each seemed to have a small globule, more or less in the center of the cell, that seemed darker and less transparent than the rest of the cell.

Others had taken notice of such things, but Brown was the first to decide that this was a common characteristic of cells, and he gave it a name. He called it the *nucleus,* from a Latin word for "a little kernel." The name was adopted, but about three-quarters of a century later it was discovered (as I described earlier) that there was a little kernel to the atom, too, and this was also named *nucleus*. The two are rarely spoken of at the same time, but if they were, they could be differentiated as the *cell nucleus* and the *atomic nucleus*. Brown discovered the cell nucleus.

The Greek word for the kernel of a nut is *karyon*. That is why cells with nuclei (and we shall see later that there are some important cells without them) are called eukaryotic cells or eukaryotes (Greek for "true nuclei").

All cells of the human body—indeed, all cells of all multicellular life—are eukaryotic. There are apparent exceptions, such as the red blood corpuscles and the platelets in the blood of human beings and other animals. These lack nuclei, but they are not really cells–not for lack of nuclei, but for the lack of essential chemical substances that nuclei contain. This is something we'll return to later.

It was impossible to see much detail in the cell, except for the shadowy nucleus itself, until chemists in the midnineteenth century began to produce synthetic dyes. It was discovered that some dyes would attach themselves to certain structures within the cell, and not to others. The cell was therefore converted into a colored pattern that yielded information that till then had not been available.

In 1879 the German biologist Walther Flemming (1843–1905) found that with certain red dyes he could stain a particular material in the cell nucleus that was distributed through it as small granules. He called this material *chromatin* (from a Greek word for "color").

When he dyed a section of growing tissue, cells were caught at different stages of cell division. The dye killed them, of course, but they served as a series of "stills" in the process, and Flemming could work out a proper order and understand what happened.

As the process of cell division begins, the chromatin coalesces into short threadlike objects that eventually came to be called *chromosomes* ("colored bodies"). Because these threadlike chromosomes seemed so notable a feature of cell division, Flemming named the process *mitosis*, from a Greek word for thread.

As cell division proceeded, the chromosomes doubled in number. They then pulled apart, half of them going to one end of the cell and half to the other. When the cell pinched together in the middle and separated into two cells, each new cell had a full number of chromosomes.

The Belgian biologist Edouard van Beneden (1846–1910) showed in 1887 that each species had cells with a characteristic number of chromosomes (in human beings the number is forty-six). Sperm cells and egg cells, however, each had half the usual species number; that is, each had a half-set of chromosomes. When a sperm cell fertilized an egg cell, the fertilized ovum that resulted had a full set of chromosomes, one half-set from the male parent and one half-set from the female parent.

It was clear that the chromosomes, or something about them, was what really controlled the characteristics of a fertilized ovum. The fertilized ovum of a rhinoceros might seem quite similar to the fertilized ovum of a cat (or a human being), but some difference in the chromosomes made one fertilized ovum capable of producing only a rhinoceros and another only a cat.

In the time of Flemming and van Beneden, no one knew just exactly how the chromosomes differed from each other, but this is a subject we will come back to later.

The protoplasm outside the nucleus (or *cytoplasm,* from a Greek word meaning "cell form") is not just a blob of gelatinous material, either. It, too, contains small structures. There are, for instance, the *mitochondria,* which were first discovered by a German biologist, C. Benda, in 1898. An average cell might contain a few hundred or even a few thousand of these structures. In Greek the name means "cartilage threads," but that is merely what they seemed to resemble as far as the discoverer was concerned. We now know that they are structures that deal with the combination of food substances and oxygen, producing energy for the use of the body.

There are also, in the cytoplasm, innumerable *ribosomes,* which were first adequately studied by the Romanian-American physiologist George Emil Palade (b. 1912) in 1956. These are tiny objects that control the synthesis of protein molecules (concerning which there will be more to say later on). There are other cell structures also, both inside and outside the nucleus, not all of which have as yet had their functions definitely determined.

The conclusion we can come to is that the cell, despite its being so tiny that it can't usually be seen with the unaided eye, is nevertheless an extraordinarily complex structure. It is not so surprising, once this is realized, that a speck of life equivalent to a fertilized ovum or a seed is capable of developing into a full-sized and very complex multicellular animal or plant.

Is it not possible, then, for a single cell to be complex enough to live independently and not merely as part of a multicellular organism?

The smallest bits of life (or potential) life known in the days before the microscope were the seeds of certain plants. Thus, when Jesus' disciples found that they could not cast devils out of a madman, Jesus explained that it was because of their lack of faith. If they had even a tiny quantity of faith, he said to them, they could do anything—even move mountains. To express the tininess of the faith necessary, Jesus said, "If ye have faith as a grain of mustard seed . . ." (Matthew 17:20). The force of this is explained by another passage, which states, ". . . a grain of mustard seed . . . which is the least of all seeds." (Matthew 13:31–32.)

(Actually, the "least of all seeds" is not the mustard seed, but the seeds of certain orchids, which weigh about a microgram—about 1/30,000,000 of an ounce—and which may just possibly be seen as tiny specks, in a good light.)

Something more astonishing than that was discovered by the Dutch microscopist Anton van Leeuwenhoek (1632–1723). From 1674 on, he spent nearly a half-century grinding tiny, but perfect, lenses (a total of 419 of them) through which he studied everything from tooth scrapings to insects.

In 1676 he focused a microscope on a drop of pond water and found it to be swarming with tiny creatures that were indisputably alive. They were no bigger than the tiniest of seeds, but they didn't just lie there, merely *potentially* alive, as seeds do. The microscopic objects van Leeuwenhoek saw were actively swimming, and there was clear evidence of internal organization. They engulfed smaller particles of life and discharged wastes. Leeuwenhoek called them "animalcules" (little animals).

We call them *microorganisms*, a term that refers to all life forms so small that they can only be conveniently studied through a microscope. (In a good light, the larger forms can, like the smallest seeds, be seen with the unaided eye as tiny specks.)

Some microorganisms move about readily by means of one or more whiplike *flagellae* (Latin for "whips") or by means of many hairlike *cilia* (Latin for "eyelashes") or by merely oozing along. These microorganisms generally lack chlorophyll and engulf their food. They are clearly tiny animals and, as a group, are called *protozoa*, from Latin words meaning "first animals."

Other microorganisms are relatively quiescent and possess, in their cells, green chloroplasts containing chlorophyll. They are tiny plants and are called *algae*.

After Schleiden and Schwann established the cell theory, it seemed that such microorganisms, unlike the larger organisms studied by those two, did not consist of cells still smaller than themselves. The German zoologist Karl Theodor Ernst von Siebold (1804–1885) made it clear in 1845 that such organisms were unicellular organisms. They consisted of single cells that were rather larger and more complex than the cells that make up parts of multicellular organisms, but they were single cells just the same.

These unicellular organisms are clearly more primitive than any multicellular organisms. It is easy to suppose that originally, before any multicellular organisms had evolved, there were *only* unicellular organisms on Earth.

However much this may seem reasonable, it remains speculative so long as we don't have observational evidence, and this would be hard to come by. If we find only dim traces of early soft-bodied creatures; how much dimmer would be the traces of microorganisms?

Yet key findings of such traces were made by the American paleontologist Elso Sterrenberg Barghoorn (b. 1915) in 1954 and afterward. He worked, to begin with, with very old rocks in southern Ontario (part of the oldest portion of North America). He shaved thin slices of these rocks and studied them under the microscope. In them he found circular structures that were about the size of unicellular animals. What's more, there were signs of smaller structures within these objects, which resembled the kind of structures within cells—including nuclei, mitochondria, and so on.

So many of these objects have by now been seen and studied that there remains no reasonable doubt that they are fossil remnants of very early eukaryotes. The earliest of these eukaryotes seem to have been a variety of algae that has been given the name of *acritarchs,* and these seem to have been up to 1,400 million years old.

It would seem that after eukaryotes came into existence, they remained the most complex form of life on Earth for 600 million years before the first and simplest multicellular organisms developed.

And yet eukaryotes, whether alone as unicellular organisms, or in combination as multicellular organisms, have only existed for the final third of Earth's existence. For the first two-thirds, there were no eukaryotes.

Might there not have been some other form of life, then, something simpler than the eukaryotes? After all, eukaryotes, even the smallest and simplest, are quite complex in structure. It is unlikely that they arose spontaneously from ordinary nonliving matter.

As it turns out, there are cells that are smaller and simpler than eukaryotes. Those are called *prokaryotic cells* or *prokaryotes*, and it is from them that eukaryotes may have evolved. Let us, therefore, consider the beginnings of prokaryotes.

# 18

## PROKARYOTES

In 1683 Leeuwenhoek, who was the first to observe micro-organisms with a microscope, noted certain objects that were just about at the limit of the resolution of his lenses. He faithfully reported them, as he did everything else he saw.

Nothing more could be done about these particularly small objects until microscopes were considerably improved. A century later, the Danish biologist Otto Friedrich Muller (1730–1784), using the better microscopes of his day, was able to study such small objects in sufficient detail to be able to detect different varieties.

Interest in these tiny objects increased sharply after the French chemist Louis Pasteur (1822–1895) was able to demonstrate in the 1860s that microorganisms are the cause of infectious disease. In 1872 the German botanist Ferdinand Julius Cohn (1828–1898) published a three-volume work on these creatures. He was the first to call them *bacteria* (from a Latin word meaning "a little rod," which rather describes the shape of some of them, though others look like little spheres and still others like tiny wriggling worms).

Bacteria are quite different from eukaryotes, or from the eukaryotic cells of multicellular organisms.

To begin with, bacteria are notable for their small size. The average eukaryotic cell is about 10 micrometers in diameter (where a micrometer is equal to one-millionth of a meter or 1/25,000 of an inch). A eukaryotic cell capable of independent life may be even larger, say 100 micrometers in diameter.

A bacterium on the other hand is only 1 or 2 micrometers in diameter, and the smallest known bacteria are only 0.1 micrometers across.

Bacteria are also notable for lacking nuclei. Since bacterial cells seem to be smaller and more primitive than the larger eukaryotic cells, bacteria are said to be prokaryotic cells, or prokaryotes, from Greek words meaning "before the nucleus." That is, they existed before the nucleus had developed.

This may seem to raise a problem. Earlier, I said that the nucleus and the chromatin material they contain are essential to cell reproduction. Thus, red blood corpuscles and platelets don't have nuclei (as I mentioned earlier) or chromatin and, in consequence, cannot grow or reproduce. That inability marks them as not being true cells. We don't run out of these blood components, however, even though they don't multiply and even though they come to the end of their existence quickly enough. Quantities of them are constantly being formed from precursor cells, which *are* cells and *do* have nuclei, and they are formed in sufficient numbers to make up for their high destruction rate.

Yet bacteria, without nuclei, manage to divide and multiply, and do so quite vigorously, too.

This is not really a puzzle. Bacteria may not have nuclei, but they *do* have the chromatin material that is necessary for growth and reproduction. This chromatin is not segregated into a nucleus as it is in eukaryotes, but is distributed throughout the bacterial cell generally. In fact, the bacterial cell is not very different in size from a eukaryotic cell's nucleus, so that a bacterium may almost be looked on as a free-living nucleus.

Bacteria also possess ribosomes so that they can manufacture protein. Those bacteria that can deal with atmospheric oxygen (a few species cannot) possess mitochondrial material.

Then, too, there are prokaryotes that possess chlorophyll as eukaryotic plant cells do. These chlorophyll-containing prokaryotes were originally called *blue-green algae* because of their color. However, once biologists recognized the importance of the eukaryote-prokaryote distinction, they couldn't help but notice that the blue-green algae were much more closely related to bacteria in structure than to ordinary algae, which are eukaryotes. At present, therefore, the blue-green algae are termed *cyanobacteria,* the *cyano-* coming from a Greek word for "blue."

It is possible that eukaryotic cells originated through a kind of combination of different varieties of prokaryotes. Mitochondria and chloroplasts both have associated with them small quantities of genetic material, which makes it tempting to suppose that they were once independent organisms.

Suppose that as prokaryotes developed, several different kinds evolved. There might be some that had well-developed flagellae for movement; some that were excellent at handling atmospheric oxygen; some that were in possession of chlorophyll. It could be that on occasion, a moving prokaryote fused somehow with an oxygen-handling prokaryote, or with a chlorophyll-containing prokaryote, or with both. These combinations might be better able to deal with the environment and to work more efficiently than any of the prokaryotes separately. They would survive and flourish.

In a way, then, we might look on eukaryotic cells as *multiprokaryotic,* just as ordinary organisms are *multieukaryotic* or, to use the more common term, *multicellular.*

(We can even imagine a further step in which organisms combine into greater wholes that can accomplish far more than a similar number of unorganized organisms could. Such organism-groups might be thought of as "societies." The insects have advanced in that direction, if we think of the well-organized populations of anthills, beehives, and termite-hills; and so of course have mammals, human societies being the best example.)

The view of eukaryotic cells as being *multiprokaryotic* is strongly upheld by the American biologist Lynn Margolis (b. 1938).

We can imagine that the combination of prokaryotes can produce larger and larger cells until we have multiprokaryotes with a thousand times the volume, and a thousand times the chromatin material, that ordinary prokaryotes have. In that case, it might become difficult to organize the process of mitosis, if we spread chromosomes all over the cell. It may be then that those multiprokaryotes survived best that collected the chromatin material into the relatively small volume of a nucleus. In this way, multikaryotes became eukaryotes.

Of course, despite the development of the eukaryotes, prokaryotes have survived to the present day and do very well. Their very simplicity and their very small size make it possible for them to grow, divide, and multiply far more quickly than eukaryotes can, and that gives them a certain advantage that the eukaryotes have lost (in order to gain other advantages). It is possible, and even likely, to be sure, that the prokaryotes of today are more advanced and complex than the original prokaryotes from which the eukaryotes evolved.

If this is all so, then prior to 1,400 million years ago, when eukaryotes made their first appearance, there must have been prokaryotes already in existence.

If the traces of the simplest eukaryotic life are hard to detect in the rocks, those of the even smaller and still simpler prokaryotic life must be still harder to detect. Nevertheless, Barghoorn and his associates have detected objects in old rocks that are of the proper size and shape to represent prokaryotic traces.

Then, too, there are a few places in the world where prokaryotes flourish and form flat, matted layers interspersed with sediment. These are called *stromatolites* (from a Greek word for "bedsheets"). The fossil remnants of such stromatolites have been found stretching back to times long before the eukaryotes.

The oldest rocks in which these prokaryote traces have been found may be as much as 3,500 million years old. This means that life existed on Earth, at least in prokaryotic form, when the planet was only a billion years old. For over 2,000 million years, more than half the time that cells of any kind have existed, earthly life consisted *only* of prokaryotes. It was a world of bacteria, with and without chlorophyll.

But even prokaryotes are complex systems, each tiny cell being filled with large numbers of different molecules, some quite complex in structure. Surely, they did not arise out of nowhere. Are there forms of life even simpler and more primitive than prokaryotes? If so, how did they come to be? What were their beginnings?

# 19

## VIRUSES

The possibilty of a simpler form of life than bacteria arose in 1880. Pasteur, who had advanced "the germ theory of disease"—that is, that all infectious disease was caused by microorganisms—was studying the disease of rabies (or, as it was at one time known, hydrophobia).

He was able to find a treatment for it, but he was *not* able to locate a microorganism that he could clearly show was the cause of the disease. He was unwilling to suppose that rabies was an infectious disease that lacked a microorganismic cause. He suggested, instead, that the microorganism in question was too small to be seen by the microscope. (The suggestion was met with a very natural skepticism.)

In 1982 a Russian botanist, Dmitri Iosifovich Ivanovsky (1864–1920), was studying tobacco mosaic, a disease of the tobacco plant that made itself evident by the formation of an unnatural mosaic pattern of the leaves. He could not find the causative microorganism for it any more than Pasteur could find it for rabies. Ivanovski mashed up infected leaves and forced the thick liquid through a very fine filter designed to remove all bacteria. If the liquid that went through did *not* infect healthy tobacco plants, then he could conclude that a bacterial cause was present but that he had simply not identified that bacterium. He found, however, that the clear liquid that passed through the filter *could* infect healthy plants.

He might have concluded from this that the microorganism that caused tobacco mosaic disease was far smaller than bacteria and could pass through a filter whose pores were too fine for bacteria. Ivanovski did not quite have Pasteur's courage, though, and chose to believe instead that his filter was imperfect and that the microorganism had passed through small cracks in it.

Three years later, in 1895, a Dutch botanist, Martinus Willem Beijerinck (1851–1931), repeated very much the same experiment, but he did not assume the filters to be imperfect. He insisted that the infective microorganism was considerably smaller than bacteria. He did not wish to speculate on its chemical or physical nature. He called it a "filtrable virus." Since *virus* is a Latin word for "poison," Beijerinck merely called it "a poison that passes through a filter."

By 1931 some forty diseases, including the common cold, measles, mumps, influenza, chicken pox, smallpox, poliomyelitis, and, of course, rabies, were known to be caused by such filtrable viruses, and *still* nothing was known about their chemical or physical nature.

In that year, however, a British bacteriologist, William Joseph Elford (1900–1942), passed a fluid containing a filtrable virus through a filter so fine that the filtrable virus was no longer filtrable. It could not pass through the tiny pores. From then on, the adjective *filtrable* was dropped, and the disease agents were simply called *viruses*.

This made it possible to estimate the size of viruses for the first time. Whereas the average bacterium was about 2 micrometers in diameter, the average virus was about 0.2 micrometers in diameter and the smallest we now know are only 0.02 micrometers in diameter. The viruses were as much smaller than prokaryotes, as prokaryotes were smaller than eukaryotes. A typical prokaryote had 1,000 times the volume of a typical virus, and a typical eukaryote had 1,000,000 times the volume.

Viruses were objects so small that there was a question as to whether they could be considered alive or not. Bacteria seemed just barely large enough to be alive; how then could an object with a thousandth its volume be alive?

In 1935, the American biochemist Wendell Meredith Stanley (1904–1971), working with a solution of tobacco mosaic virus, put it through a series of procedures that had recently succeeded in producing crystals of protein molecules. He obtained fine needlelike crystals of tobacco mosaic virus. These crystals, when separated, dried, and then dissolved in fresh water, showed all the properties of the virus and could infect healthy tobacco plants.

This seemed to speak in favor of a virus being a nonliving protein molecule, for it seemed unthinkable that a living organism could exist in crystalline form. Yet, on the other hand, the virus could reproduce itself once it was inside a cell and it could, apparently, manage to get inside such a cell in the first place. That seemed to be an ability only a living thing could have. If they crystallized, it might be that, even though they were alive, they were so simple in structure as to possess the crystallizing properties of a protein molecule.

And yet were they only proteins? Chemical tests of viruses clearly showed the presence of protein, but might there be something else there in addition?

In 1936 two British biochemists, Frederick Charles Bawden (b. 1908) and Norman Wingate Pirie (b. 1907), showed that tobacco mosaic virus was only 94 percent protein. The remaining 6 percent was a substance called *nucleic acid*.

Nucleic acid had been discovered in pus in 1869 by a Swiss biochemist, Johann Friedrich Miescher (1844–1895). He called it ''nuclein'' because it seemed to be associated with cell nuclei. Because it was later found to exhibit acid properties, the name was changed to nucleic acid.

It took three-quarters of a century to work out the

structure of nucleic acid in all its details, but by the time
Bawden and Pirie made their discovery, the structure of
nucleic acid was understood. There were two major varieties, *ribonucleic acid* and *deoxyribonucleic acid,* usually
abbreviated as RNA and DNA respectively. When either
of these existed in combination with a protein, the two
together formed a *nucleoprotein.*

As it turned out, all viruses were found to be nucleopro-
teins in nature. In the case of tobacco mosaic virus and a
number of others, the nucleic acid involved was RNA. In
the case of still other viruses, it was DNA.

Nucleic acids also existed in cells, of course, for it was
there that they were discovered. The German biochemist
Robert Joachim Feulgen (1884–1955), using staining reactions he had devised himself, showed in 1923 that DNA
was strongly concentrated in the nucleus of a cell, while
RNA existed in the cytoplasm.

The Swedish biochemist Torbjorn Oskar Caspersson (b.
1910) studied the nucleic acids in the cell in still greater
detail and, in the mid-1930s, made it quite clear that DNA
existed not merely in the nucleus, but specifically in the
chromosomes.

After that, it was possible to think that just as a bacterium could be viewed as a kind of isolated cell nucleus, a
virus could be viewed as an isolated cell chromosome.

Chromosomes had, by then, earned a position of pecu-
liar importance to life in the eyes of biologists. In 1865 the
Austrian botanist Gregor Johann Mendel (1822–1884) had
worked out the mechanism of heredity, the manner in
which physical characteristics passed from parent organisms to children. For this he had to postulate the existence,
within the organism, of certain factors of inheritance that
behaved in particular ways.

Mendel's work was neglected for many years, but it was
brought to the general attention of biologists in 1900 by a
Dutch botanist, Hugo Marie de Vries (1848–1935). By
that time, much more was known about the details of
cellular structure, and in 1902 an American biologist,
Walter Stanborough Sutton (1877–1916), pointed out that

chromosomes, in the course of cell division, behave precisely in the manner that Mendel's factors of inheritance would be expected to behave.

The chromosomes, therefore, turned out to be the carriers of heredity and must in some way control the chemistry of the cell in order that that cell and the organism of which it was a part could display various characteristics as inherited from the parents. In fact, it was not the characteristics themselves that were inherited but the chromosome that produced the characteristics.

The Danish botanist Wilhelm Ludwig Johannsen (1857–1927) realized that the chromosomes were far too few to control all physical characteristics if each chromosome was supposed to control but one. He therefore suggested in 1909 that the chromosomes were divided into small sections, each of which gave rise to a single characteristic. These small sections he called *genes*, from a Greek word meaning "to give rise to."

When a virus invades a cell, then, it is a foreign and parasitic chromosome that can make use of the cell machinery for its own purposes; that is, for manufacturing more viruses like itself. Some viruses are moderate in their action and parasitize a cell without killing it. Other viruses kill the cell in the process of their own exuberant multiplication.

Since before multicellular organisms arose, life on Earth consisted of unicellular organisms only, and since before eukaryotic cells arose, life on Earth consisted of prokaryotes only, might it be that before the existence of any cells at all, life on Earth consisted only of viruses?

We have no indication of any kind, as yet, that this was, in fact, really so. We can be sure that if viruses existed before cells, they were not the viruses of today. All known viruses now in existence are parasites on cells and cannot multiply except by using the machinery within already existing cells. It may even be that the viruses of today have evolved in a "degenerative" way, from cells. That is, they are cells that lost some of their chemical capacities precisely because it was so much easier to allow more independent cells to do the job for them.

There are rickettsial cells, for instance, or, as they are usually called, *rickettsia*. These were first discovered by the American physician Howard Taylor Ricketts (1871–1910), who found, in 1909, such cells to be the causative agent for Rocky Mountain Spotted Fever. Rickettsia are like small bacteria that cannot live independently because they lack certain proteins called *enzymes* that catalyze (that is, bring about rapidly) key reactions of life. Rickettsia can only grow and multiply if, inside the cells they infest, they can find and make use of the enzymes they are missing.

There are viruses that are smaller than rickettsia but that are still rather complex, and a series of progressively smaller and simpler viruses, all of which lack more and more of what is required for independent life. The smallest viruses retain only the bare ability to penetrate into a cell, and once there they multiply entirely by the control they exert over a cell's enzymes, contributing virtually no enzymes of their own.

And yet, since it seems unlikely that even the least complex bacterial cell could have arisen at a bound without simpler precursors, we can only assume that prokaryotes were generally preceded by viruslike objects capable of some form of independent life. Little by little, in the course of the first billion years or so of Earth's existence, these viruslike objects developed to the point of being bits of life complex enough for us to recognize as prokaryotes.

These precursors of life must have formed out of simple molecules of the type found all about us in the air and the ocean. Therefore, before we can speculate further on the beginnings of life, let us consider the beginnings of Earth's ocean and atmosphere.

# 20

## OCEAN AND ATMOSPHERE

Earlier in this book, I described the manner in which the Babylonians and their predecessors explained the origin of the Earth: the conversion of the Chaos of illimitable ocean into the order, or Cosmos, that marks the present Universe. The Jews, during the Babylonian captivity, picked up elements of this tale, which then appeared in the first chapter of the Book of Genesis.

The Book of Genesis begins, "In the beginning God created the heaven and the earth" (Genesis 1:1), and then goes on to give the details.

To begin with, "the earth was without form, and void; and darkness was upon the face of the deep." (Genesis 1:2). The "void" and the "deep" are both words that denote the original chaos that is "without form." Chaos can be visualized as a kind of hectic ocean in which all the substances that go into the making of the Universe exist in random, disorderly mixture.

However, "the spirit of God moved upon the face of the waters" (Genesis 1:2), and the will of God imposed order upon it by setting up a series of separations. On the first day, he separated light from darkness, creating day and night. On the second day, he created the sky to separate the waters beneath (the ocean) from the waters above (the rain). And, on the third day, he separated the water from the land, thus creating not only the continents but the ocean as we know it today.

Thus, in the biblical view, the ocean has existed as it is now from the third day of creation.

The ocean, however, can at least be seen. The air is invisible and we know it is there only because its motion can be felt as a wind. It can easily be ignored, and as a matter of fact, the Bible doesn't bother describing the creation of the atmosphere. Perhaps its creation can be omitted because, in a way, air can be viewed as chaos since there is no visible order to it. Perhaps, it is just a bit of chaos left over from the beginning and required no creation.

The air, until modern times, was presumed to extend upward in more or less the same condition in which it exists at sea level, until it reached the sky, which the ancients (and the Bible) assumed to be a solid vault. To be sure, the thought of the air reaching the sky was not very remarkable, since the sky, to most people in early times, was not thought to be very high. It might just clear the mountains. Thus, in a Greek myth, the Titan Atlas, as punishment for having warred against Zeus, was condemned to hold up the sky. At one point, the human hero Hercules stood on a mountain peak and was then tall enough to take over the task for a short while.

To the ancients, water and air were two of the *elements*, or fundamental substances, making up the Universe. There was a tendency to think of all liquids as owing their liquidity to an admixture of water, and of all vapors as owing their vaporousness to an admixture of air.

The first person to recognize clearly that there were airlike substances that were quite distinct in properties from air was a Flemish physician, Jan Baptista van Helmont (1580–1644). He coined a word in 1624 for any kind of vapor that possessed airlike qualities and called each a *gas*.

This was a final echo of the thought of air and ocean as forms of chaos, for "gas" is but a simplified spelling of "chaos."

Van Helmont's invented term was largely ignored at first, and for a century and a half after the coinage, chemists still talked of the gases they discovered and worked with as types of air. There was "fixed air" and "fire air" and "phlogisticated air" and "dephlogisticated air," and so on. It was the French chemist Antoine Laurent Lavoisier (1743–1794) who rescued the term and fixed it into the chemists' and the world's vocabulary.

Meanwhile, though, a discovery had been made that fundamentally changed all views concerning the air. The Italian physicist Evangelista Torricelli (1608–1647) had, in 1643, succeeded in balancing a column of air against a column of mercury and had shown in this way that air had weight and that it pressed down on every square inch of surface (including the surface of the human body) with a weight of 14.7 pounds (6.7 kilograms). Human beings are unaware of this weight because the fluid contents of the body press outward in all directions with a balancing force.

This meant that air could not fill the Universe to indefinite heights. In fact, from its weight, one could calculate that if air had the same density throughout, then it would only be about 5 miles (8 kilometers) in height.

This was not so, since, as the British physicist Robert Boyle (1627–1691) showed in 1662, air is compressed by pressure. This means that air at sea level is weighed down by the air at higher levels and is packed together more tightly, and is denser, as a result. As one climbs up the slope of a mountain, the air one encounters has less air above it, so that it is under less pressure. This means it grows less dense, thins out, and takes up more room. Therefore, air stretches upward to heights far greater than five miles, though at the cost of becoming thinner and thinner, wispier and wispier.

Air becomes too thin to support human life at about 6 miles (9.6 kilometers) above sea level, is reduced to traces at 100 miles (160 kilometers), and is virtually undetectable at 1000 miles (1600 kilometers). This means that Earth's airy envelope, or *atmosphere* (from Greek words meaning "ball of vapor"), is restricted to the immediate neighborhood of the Earth.

This, in turn, means that the vast spaces between astronomical bodies—between the Earth and the Moon, for instance—contain nothing except all but imperceptible traces of matter and may be considered a *vacuum* (from a Latin word for "empty").

Ordinarily, it is human experience that gases such as air expand to fill all the space available, yet Earth's atmosphere shows no perceptible tendency to expand outward into the vacuum.

The reason for this is that the atmosphere is held closely to Earth's surface by the action of gravity, a force first explained in satisfactory fashion by the English scientist Isaac Newton (1642–1727) in 1687. An object can escape from a gravitational pull if it moves fast enough (the *escape velocity*), but the escape velocity from Earth is 7 miles (11.25 kilometers) per second, and the air, or any sizable portion of it, rarely moves at more than 1/100 that speed in even the most violent tornado.

The atmosphere, however, like all other parts of the Universe, is made up of tiny *atoms* that, in turn, may exist in groups called *molecules*. In solid matter (and, to a far lesser extent, in liquids) the molecules are bound together and cannot move separately. In gases such as air, on the other hand, the molecules barely influence each other and each one moves more or less independently of the rest.

In the 1860s, a Scottish mathematician, James Clerk Maxwell (1831–1879), worked out the *kinetic theory of heat*, which showed the speeds at which different atoms or molecules move. As the temperature rises, the average speed of motion rises, too. There is, however, always a range. At any temperature, there are always some molecules moving faster (a few much faster) and some molecules moving slower (a few much slower) than the average.

This means that in any atmosphere there is always the

chance that occasional molecules will be moving fast enough to escape into the surrounding vacuum, if those molecules happen to be in the upper atmosphere and can reach the vacuum without striking another molecule and losing some of their speed. Every atmosphere "leaks," in other words. In Earth's case, this leak is so slow that even after billions of years no perceptible amount of the atmosphere has been lost.

The smaller the astronomical body and the weaker its gravitational pull, the lower its escape velocity and the greater the chance that individual molecules will have enough speed to escape. In short, the smaller and less massive the body, the faster the atmosphere will leak.

Again, the hotter the astronomical body, the faster the individual molecules of the atmosphere will move, and the faster it will leak. Finally, the smaller a molecule, the faster it will move at a given temperature. In any atmosphere, therefore, smaller molecules will leak away faster than larger ones.

If a heavenly body is small enough, or hot enough, or both, then any atmosphere that may have existed at one time will have leaked away in a comparatively short period and it will be airless. If it is large enough, or cool enough, or both, then it will have an atmosphere.

Thus, the eight most massive bodies of the solar system all have substantial atmospheres. These are, in decreasing order of mass: the Sun (which has an atmosphere despite its ferocious surface temperature of nearly 6000° C.), Jupiter, Saturn, Neptune, Uranus, Earth, Venus (despite its surface temperature of 475° C., which is far above the boiling point of water), and Mars.

Mars, to be sure, has only a thin atmosphere, about 1/100 the density of that of Earth. The ninth most massive body, Mercury, is too small for an atmosphere, especially since it is so close to the Sun as to have a high surface temperature, though not as high as that of Venus.

The tenth most massive body is Ganymede, the largest satellite of Jupiter. It has no atmosphere either, even though it is much cooler than Mercury is. The eleventh most massive body, Titan, which is the largest satellite of Saturn, is somewhat smaller than Ganymede, but it is much cooler still and therefore can and does hold an atmosphere. The twelfth most massive body, Callisto, the second largest satellite of Jupiter, has no atmosphere. The thirteenth most massive body, Triton, the larger satellite of Neptune, is so cold that it may have an atmosphere. We don't know yet.

All the myriad objects in the solar system that are less massive than Triton do not have atmospheres.

So far, then, Earth does not seem to be unique in having an atmosphere, since eight other bodies in the solar system, and possibly nine, have one. However, we will return to this point shortly and show where the uniqueness lies.

In the case of liquids, we find that while the molecules of which they are composed are bound together, the binding is not as tight as it is in the case of solids. There is a far more perceptible tendency for individual molecules to break away from the body of a liquid than from the body of a solid, all other things being equal. Liquids, in other words, have a tendency to vaporize and become gaseous in form. Thus, water has a tendency to turn into water vapor.

We can observe this after a rain, when the moisture in the streets gradually disappears. All open bodies of water, even the ocean, are continually vaporizing, so that one of the constituents of the atmosphere is water vapor. The water vapor content of the atmosphere does not build up indefinitely, for there is also a tendency for the vapor to condense back into liquid water. The evaporation is balanced by precipitation and, between the two, the water vapor content of the atmosphere remains reasonably constant in the world as a whole.

Because there is always water vapor in the air, the water molecules in that vapor may occasionally rise into the upper atmosphere, and if they are then moving quickly enough and don't lose the speed through collisions, they may escape. On Earth, the leakage is insignificant even over billions of years, but on worlds where the leakage is rapid any supply of liquid water would dwindle away and the world would turn dry.

Thus, the Moon and Mercury are totally dry. Venus has a totally dry surface, too, because of its high surface temperature, but there is some water vapor in its cold upper atmosphere still.

If the temperature is below 0° C., water exists in the solid form as ice, which evaporates much more slowly than liquid water does. This means that all the worlds that remain farther from the Sun than Earth does, through all or most of their orbits, can retain water even if they are fairly small—but only as ice.

Thus, Mars has a small water supply—as ice. Most of the satellites of the outer planets, together with some asteroids and almost all comets, are icy. There is some reason to think that Europa, the smallest of Jupiter's four large satellites, is covered with a world-girdling ocean of liquid water, but if so, the ocean is, in turn, covered with a perpetual world-girdling layer of ice. In the case of the four giant planets, Jupiter, Saturn, Uranus and Neptune, water probably makes up only a small percentage of their surface materials.

There you have the uniqueness of Earth's ocean. Earth is the only world in the Solar system to have a broad liquid expanse of surface water uncovered by ice.

This is important. In a gas, molecules are separated by comparatively large distances, and chemical reactions, which depend on collisions between molecules, may not proceed at the speed and with the variety that one requires in a living system. In a solid, molecules are in virtual contact, but they cannot move freely and that cuts down the speed and variety of chemical reactions. In a liquid, molecules are also in virtual contact, but they can move about far more freely than in solids. A liquid is, therefore, the ideal medium in which we might expect life to begin.

What's more, of various liquids, water is particularly suitable because it has a high solvent ability and can carry a variety of substances in solution. Molecules, which would ordinarily be part of a solid if left to themselves, behave as if they were part of a liquid, when in solution.

Life is usually considered to have begun in the ocean, and the fact that Earth has millions of cubic miles of liquid water exposed to the rays of the Sun (a natural and copious energy source) makes the world an ideal place for life to begin. The fact that Earth is the only world in the Solar system of which this may be said might cause one to suspect that life does not exist elsewhere in the solar system.

(There is a chance, of course, that life is possible on a completely different basis from that which we know on Earth, so that life of some sort might exist on a planet whose environment we would consider irredeemably hostile to life. There is, however, no evidence whatever that this might be so, at least so far, and until such evidence is found, it would be dangerous to consider nonaqueous life as anything more than interesting speculation.)

But let us return to the atmosphere—

As I have said, eight and possibly nine worlds in the Solar system have atmospheres, but have we any right to assume that all the atmospheres are the same in nature?

Until modern times, there was a general assumption that air was an element, a unitary substance, all parts of which were alike. It was not supposed to be a mixture or a combination of different substances. If that were so, it might seem natural to suppose that the same air that existed here would exist on any other world that had an atmosphere.

The assumption, however, is wrong.

Beginning with van Helmont, chemists worked with a number of vapors of different properties, but they were produced in the laboratory under specialized conditions and no one assumed they existed in the air. After all, there are many liquids that are not water—alcohol, turpentine, mercury, olive oil, and so on—and chemists were aware of them in early times. Still, no one thought that these liquids were to be found in the ocean. At best, if they existed there, it would be as insignificant impurities. There was salt in the ocean, of course, but that was only a dissolved solid. The only liquid making up the water of the ocean was—water.

In the same way, there might be dust in the air, or whiffs of water vapor, or other odorous vapors of one sort or another, but these were only insignificant impurities. In essence, air was simply—air.

In 1754 a Scottish chemist, Joseph Black (1728–1799), studied the gas we now call *carbon dioxide*. Black showed that what we now call *calcium carbonate* lost carbon dioxide on heating, becoming *calcium oxide*. This was the first indication that you could produce a gas by simply heating a solid.

Black also showed that if you bathed calcium oxide in carbon dioxide, it changed back to calcium carbonate. In addition, if you simply allowed calcium oxide to stand in air, it would very slowly change to calcium carbonate. This meant there had to be carbon dioxide in the atmosphere as a natural component of air.

However, the carbon dioxide was again merely a trifling impurity. We now know that it only makes up 0.035 percent of the air. There is far less of it in air than there is of water vapor.

Black was also interested in the fact that although a candle can burn indefinitely in air, it can only do so if it is in open air. If it is made to burn inside a closed vessel so that there is only a limited supply of air, the candle eventually goes out even though there is still unburned wax remaining and though there is still air within the vessel.

Black knew that the burning candle produced carbon dioxide and that nothing would burn in carbon dioxide. A flame plunged into a container of carbon dioxide would go out. However, if Black added chemicals that absorbed carbon dioxide as fast as it formed, a candle would still go out if the air supply was limited and even though air remained that did *not* contain carbon dioxide.

Black passed the problem to a student of his, the Scottish chemist Daniel Rutherford (1749–1819). Rutherford repeated the experiments very carefully and, in 1772, obtained and studied a sample of a gas that was not carbon dioxide and yet in which candles wouldn't burn and mice quickly died. It is the gas we now call *nitrogen*.

In 1774 the English chemist Joseph Priestley (1733–1804) isolated a gas that was of just the opposite nature. A smoldering splint of wood would burst into flame if placed in this gas, and mice behaved very friskily in it. Priestley himself enjoyed breathing it. It was the gas we now call *oxygen*.

Finally, in 1778, Lavoisier, who had popularized the term *gas,* performed a series of experiments that made it quite clear that air was not an element, but a mixture of two different gases, nitrogen and oxygen, in about a 4:1 ratio by volume. We now know that nitrogen makes up 78 percent of the volume of air, and oxygen 21 percent.

This adds up to 99 percent but almost all the remaining percentage is *argon,* a gas that was first discovered in 1894 by the English physicist John William Strutt, Lord Rayleigh (1842–1919) in collaboration with the Scottish chemist William Ramsay (1852–1916).

Then there is carbon dioxide in a tiny amount, still other gases in even tinier amounts, and, of course, water vapor, which in any particular sample of air varies somewhat in quantity.

Now we can see where the uniqueness of the Earth's atmosphere lies: The Earth is the only world in the solar system that has an atmosphere in which oxygen is a major constituent.

This requires explanation.

The uniqueness of a liquid water ocean is easy to understand, since it depends upon temperature. On a world that is too hot, water boils away and exists only as vapor; on a world that is too cold, water freezes permanently into ice. Earth is the only world in the Solar system in which the temperature is in the right range to produce liquid water and the gravitational pull is strong enough to hold it.

An atmosphere that is unique in possessing oxygen is not so easy to explain. Oxygen could easily exist in an atmosphere as hot as Venus's or as cold as Titan's, if temperature were the only consideration. Yet it doesn't. It doesn't exist as a free gas on any other world in any quantity—only on Earth.

The puzzle is, why does it appear in Earth's atmosphere?

Oxygen is a very active gas. That is, it easily combines with other substances. Left to itself it would gradually combine with various substances in the Earth's crust and eventually disappear.

As a matter of fact, human beings have, for at least half a million years (and particularly in the last century), been burning wood and other fuels. In the process of combustion, the hydrogen and carbon atoms in these fuels are combining with oxygen in the air. The hydrogen combines to form water molecules and the carbon combines to form carbon dioxide molecules. For that matter, we and almost all other forms of life obtain energy by the combining of carbon and hydrogen atoms in the food we eat, or in our tissues, with oxygen of the air.

For all these reasons, we should expect to see the oxygen content of the atmosphere decrease steadily, year by year, until, in not very long a time, our kind of life would come to an end. Yet this doesn't happen—and the percentage of oxygen in our atmosphere remains constant for year after year. The only way of explaining this is to suppose that oxygen is formed continuously on this planet at a rate that balances its consumption. But how?

An answer to the question began to appear when Priestley, who was soon to discover oxygen, experimented in 1771 with air in which a candle had burned itself out so that nothing more would burn in it. The gas that remained in the container was now made up of nitrogen and carbon dioxide only, and a mouse placed in it died almost at once. In order to check whether the mixture was fatal to all life, Priestley put a sprig of mint into a small container of water and put that into a jar of the burned-out air.

To his surprise, the mint did not die. In fact, it seemed to flourish. After some months, during which the sprig continued to live and grow, he put another mouse into what had been dead air—and it lived. In fact, a candle would now burn in it again.

What it seemed to mean was that what animals and combustion consumed, plant life restored. In other words, animals (and burning fuel) combine food or fuel with oxygen and produce carbon dioxide and water; plants consume carbon dioxide and water and produce oxygen and the carbon/hydrogen substances of their tissues. The two tendencies remain in balance.

Natural change always produces energy. To reverse a natural change requires an *input* of energy. The natural change is from carbon and hydrogen plus oxygen to carbon dioxide and water. That produces the energy life uses for its purposes. When plants, however, convert carbon dioxide and water into their tissues plus oxygen, that reverses the natural change and requires an input of energy. Where do the plants get the energy for this purpose?

In 1779 a Dutch physician, Jan Ingenhousz (1730–1799), showed that plants only produce oxygen in the sunlight. It takes solar energy, then, to enable plants to reverse the natural change and build up their own tissues (to serve as food and fuel for animals, including human beings). The process is, for that reason, called *photosynthesis*, from Greek words meaning "to build up by light."

This explains why Earth has an atmosphere containing a great deal of such an active gas as oxygen, and why the oxygen doesn't combine with other substances and simply disappear. Earth has a flourishing system of life, including plants that produce oxygen as fast as it disappears.

This must mean that the other worlds in the solar system that possess atmospheres without oxygen must lack that gas because they do not possess a flourishing system of life. At least, they don't possess a flourishing system of *our* kind of life, and, so far, we lack any evidence that any other kind exists or is even possible.

It means something else, too. In the days when life was just beginning to form on Earth, there was no life already existing. If there was no life on Earth, there could be nothing but, at most, traces of oxygen in its atmosphere.

We conclude, then, that life formed when Earth's atmosphere lacked oxygen.

What was Earth's atmosphere like at that time, then?

We can reach some conslusions about this by considering the types of atoms in the Universe that might contribute toward the making of an atmosphere. The dozen most common atoms present in the Universe (according to present-day astronomical evidence) are, in order of decreasing abundance, hydrogen (H), helium (He), oxygen (O), neon (Ne), nitrogen (N), carbon (C), silicon (Si), magnesium (Mg), iron (Fe), sulfur (S), argon (Ar), and aluminum (Al).

Hydrogen atoms, the simplest of all, make up 90 percent of all the atoms in the Universe, while helium atoms, the next simplest, make up another 9 percent. The other ten types of atoms, taken together, make up almost all the remaining 1 percent. We can ignore everything else since there are simply not enough of the atoms of other varieties than these twelve to be of major importance in the structure of a planet or its atmosphere.

Of the twelve types of atoms I have listed, four—silicon, magnesium, iron, and aluminum—form only combinations that are solid and cannot contribute to an atmosphere.

Of those that remain, three—helium, neon, and argon—do not form any combinations but remain single atoms. Conglomerations of such atoms are gases and can contribute to atmospheres.

Of the last five, oxgyen, in the presence of a vast oversupply of hydrogen, will combine with hydrogen to form water molecules, each consisting of two hydrogen atoms and an oxygen atom ($H_2O$); nitrogen will combine with hydrogen to form ammonia molecules, each consisting of three hydrogen atoms and a nitrogen atom ($NH_3$); carbon will combine with hydrogen to form methane molecules, each consisting of four hydrogen atoms and a carbon atom ($CH_4$), and sulfur will combine with hydrogen to form hydrogen sulfide molecules, each consisting of two hydrogen atoms and a sulfur atom ($H_2S$). Even after all the oxygen, nitrogen, carbon, and sulfur combine with hydrogen, an overwhelming number of hydrogen atoms remain, and these combine with each other to form hydrogen molecules made up of two hydrogen atoms each ($H_2$).

These last substances are all gaseous at ordinary temperatures except for water, which is a liquid but one that is easily turned into a vapor. There are therefore eight gases and one liquid that can contribute significantly to an atmosphere. In order of decreasing abundance, they are hydrogen, helium, water, neon, ammonia, methane, hydrogen sulfide, and argon.

Any astronomical object that is large enough to have a gravitational field capable of holding all of these substances will be formed almost entirely of hydrogen and helium, and its atmosphere will consist of those two substances plus very minor quantities of other gases. This is true of the Sun, for instance, whose enormous gravitational field can hold even hydrogen and helium, the smallest atoms, and can do so even at the high temperatures of the Sun's surface.

It does not hold it absolutely, however. The Sun's electrical activity, in the form of flares, can break up atoms into negatively charged electrons and positively charged nuclei. The nuclei, which are the more massive and therefore more significant, shoot out from the Sun in all directions and make themselves felt far out into the planetary system.

These speeding, electrically charged particles make up the *solar wind*. The existence of the solar wind was only understood in the 1950s, when rocket exploration of space began, and it was given its name in 1958 by the American physicist Eugene Newman Parker (b. 1927). The Sun loses only an insignificant fraction of its mass to the solar wind, but it plays an important role in the mechanics of the solar system.

Bodies considerably smaller than the Sun may also retain hydrogen and helium and have them make up almost the whole of their atmosphere, provided they are considerably cooler than the Sun. The outer planets are massive enough and cold enough, at their surface, to retain these gases. Indeed, it was because they were comparatively cold when they were forming and could hold these abundant gases that they grew so large. Their large size increased their gravitational pull and made it even easier to collect more of the gases. This "snowball effect" produced the giant planets, Jupiter, Saturn, Uranus, and Neptune, all of which have hydrogen-helium atmospheres.

But what about the planets that are relatively close to the Sun? They were much warmer than the outer planets and could not hold on to the tiny atoms of hydrogen and helium in any but a trifling way. For the most part, they were built up of silicon, magnesium, iron, aluminum, and other even less common elements capable of forming metallic or stony solids that could hang together through the force of chemical bonding and that do not depend on gravitational pull to preserve their integrity. Because these elements are comparatively rare, the planets near the Sun are much smaller than the outer giants.

If a warm planet is not *too* small, it may hold some of the common gaseous substance because the atoms and molecules of those gases may have a tendency to combine with some of the rocky or metallic solids, more or less loosely, and to be trapped inside the forming planet. Helium, neon and argon formed no combinations at all and they escaped more easily than the others so that Earth today has only very small quantities of these gases in its atmosphere. Very little of the gaseous hydrogen would have been trapped, too. These light gases that the Earth's gravitational pull had not been able to collect were swept far into the outer reaches of the solar system by the solar wind, and there some at least were collected by the giant planets.

As the Earth squeezed together in the process of formation and grew more compact, the liquid and gaseous substances were forced outward. Water molecules squeezed out and formed an ocean in the lower-lying basins. Ammonia, methane, plus a little hydrogen sulfide, squeezed out to form the atmosphere and to it was added water vapor.

These molecules were large enough to be held by Earth's gravitational pull.

We might call Earth's resulting atmosphere of ammonia, methane, water vapor, plus a little hydrogen sulfide, Atmosphere I. It may not have stayed very long for it was probably unstable so close to the Sun. The water molecules penetrating to the upper atmosphere would be broken up by the ultraviolet light of the Sun. (This is called *photolysis*, from Greek words meaning "breaking up by light.")

The water molecules would break up into their constituent atoms of hydrogen and oxygen. The Earth's gravitational field would not hold hydrogen, which would leak away, but it would hold oxygen.

Oxygen, however, is chemically active. It pulls the hydrogen atoms away from the ammonia molecules, re-forming water, while the nitrogen is left to itself. Nitrogen atoms are not active. They tend merely to double up, forming nitrogen molecules made up of two nitrogen atoms ($N_2$).

The oxygen also pulls hydrogen atoms away from the methane molecules, re-forming water and combining with the carbon atoms to form carbon dioxide, with molecules made up of one carbon atom and two oxygen atoms ($CO_2$). The oxygen pulls the hydrogen atoms away from hydrogen sulfide, re-forming water and combining with sulfur to form sulfur dioxide, with molecules made up of one sulfur atom and two oxygen atoms ($SO_2$).

Both the carbon dioxide and the sulfur dioxide could combine with the rocky material of Earth's solid crust and could also dissolve in Earth's ocean. The sulfur dioxide could in this way be removed from the atmosphere in all but traces. The much more common carbon dioxide would remain in the atmosphere in substantial amounts.

The result of all these changes would be that the atmosphere is converted to one made up of nitrogen and carbon dioxide, plus water vapor. This might be called Earth's Atmosphere II.

Besides the Sun and the four giant planets, which all

have hydrogen/helium atmospheres, there are four worlds in the solar system with atmospheres: Venus, Mars, Titan, and Earth.

Of these, Venus and Mars both have nitrogen/carbon dioxide atmospheres. Titan, which is much farther from the Sun than either of those two inner planets, and where the Sun's ultraviolet light is far less concentrated, is only partway there. Its atmosphere is nitrogen/methane.

On Earth, life began while it possessed Atmosphere I or Atmosphere II (or, perhaps, during the transitional stage between). Once life began, there soon developed a new way of forming oxygen, far more rapid and efficient than the method of photolysis. This new way, photosynthesis, produced oxygen at the expense of carbon dioxide so that, eventually, Earth (alone among the planets) had a nitrogen/oxygen atmosphere, which we might call Atmosphere III.

Let us, then, at this point, return to the question of the beginnings of life.

# 21

## LIFE

We have carried life back to its simplest known form—the virus—and found that to consist of nucleoprotein; that is, an association of nucleic acid and protein. If we are now to move still further back, toward the beginnings of life of any kind, we must consider these two types of substances. Let's start with protein.

Prior to modern times, there was the tendency to think of food as food. Foods differed in taste, but that might be viewed as a purely subjective matter. In a pinch, it seemed, any kind of food that didn't actually contain poison would keep one going.

It was in 1815 that this was shown to be wrong. France had gone through a revolution and a quarter century of wars—the plight of the poor was terrible. A French physiologist, Francois Magendie (1783–1855), undertook the task of determining whether a nourishing food could be obtained from gelatin, which could be derived cheaply from otherwise all but unusable cuts of meat.

He found the answer to be in the negative. Life could

not be sustained on gelatin alone. Clearly, some foods were better than other foods.

This inspired considerable research into the different components of foodstuffs, and in 1827 an English chemist, William Prout (1785–1850), divided food into three main components: fats, carbohydrates, and what were then called "albuminous substances." (It was called that because it was found in egg white or *albumen*, from a Latin word for "white.")

Of these three types of substances, fats and carbohydrates were made up of only carbon, hydrogen, and oxygen atoms. The albuminous substances contained these three plus nitrogen and, sometimes, sulfur. What's more, the albuminous substances seemed to be more complex and variable in chemical structure than the other two were.

A Dutch chemist, Gerardus Johannes Mulder (1802–1880), studied the chemical structure of the albuminous substances and, in 1838, concluded that they were built up of a basic building block to which various amounts of modifying structures were added. He called the basic building block *protein* from the Greek word for "first," because it was out of these building blocks that the albuminous substances were built up. Mulder's speculation turned out to be not quite right, but the name remained and came to be applied to the albuminous substances as a whole. These have been known as proteins ever since.

Continued studies of protein molecules showed that they were *polymeric molecules* or *polymers*, from Greek words meaning "many parts." The name is applied to any giant molecule that is built up of small units (or "parts") hooked together. Starch and cellulose are polymeric molecules built up of many units of *glucose*, a simple sugar. Rubber is a polymeric molecule built up of many units of a simple hydrocarbon (made up of hydrogen and carbon atoms only) called *isoprene*. Modern plastics and synthetic fibers are polymeric molecules built up of some simple unit or other.

In most polymers, there is only one unit, repeated over and over again. Sometimes there are two different units that occur alternately along the chain. Very rarely are there more than two units involved in building up a polymer.

The protein molecules, it turned out, are built up of units called *amino acids,* which contain atoms of carbon, hydrogen, oxygen, and nitrogen (plus, occasionally, sulfur). What makes proteins quite different from other polymers is that the amino acids out of which protein molecules are built up occur in *twenty* different varieties. Any protein molecule is likely to have some of each variety as part of its structure.

During a period of more than a century, these amino acids were isolated from various proteins, and their structures determined. The first amino acid to be studied was isolated in 1820 by a French scientist, Henri Bracconot (1781–1855). The last was *threonine,* which was isolated by the American biochemist William Cumming Rose (1887–1985) in 1935.

This large number of different amino-acid units is of importance. They can be arranged in any order, and every different order produces a molecule with its own characteristic properties. If we start with only one of each of the twenty, these will suffice to form (believe it or not) about two and a half billion billion different orders and, therefore, different molecules.

Suppose we consider the hemoglobin molecule (located in our red blood corpuscles and serving to carry oxygen from the lungs to all the cells of the body). It contains 539 amino acids, including a sizable number of each of the twenty varieties. The number of different arrangements in which we can place those hundreds of amino acids is equivalent to a 1 followed by 620 zeroes. The number of all the subatomic particles in the entire known Universe is virtually zero compared to this mighty number. Yet for hemoglobin to work properly only one arrangement is wanted. A mistake in a single amino acid in hemoglobin can produce a molecule that works with dangerous imperfection.

Most of the proteins that were first studied were not particularly notable for their services to life. They were largely structural in nature: *keratin,* in hair, nails, hooves, claws, skin, and feathers; *collagen,* in tendons and connective tissue, and so on. Such proteins did not vary very much from individual to individual, or even from species to species.

What seemed to be much more "lifelike" were what were originally called *ferments.* These were known from prehistoric times, since yeast fermented fruit juices, soaked grain, and dough to produce alcohol and bubbles of gas and to leave behind wine, beer, and soft bread.

By the early nineteenth century, it was understood that there were ferments in living tissue, substances that in very small quantities could bring about certain specific, rapid chemical changes, changes that would proceed only very slowly in the absence of those ferments. This is an example of something generally referred to as *catalysis.*

The first ferment to be isolated and studied was *diastase.* The French chemist Anselme Payen (1795–1871) obtained it from grain and found that it brought about, or catalyzed, the speedy breakdown of starch to sugar.

A year later, Schwann (one of the founders of the cell theory) isolated the first animal ferment. It came from the stomach lining, and he called it *pepsin* from a Greek word meaning "to digest," because it catalyzed the breakdown of protein molecules to smaller fragments.

In 1876 the German physiologist Wilhelm Kühne (1837–1900) suggested that *ferment* be restricted to those catalysts that worked only in living cells. Those which could be isolated and made to work outside cells should, he felt, be called *enzymes,* from Greek words meaning "in yeast," since they acted outside cells as ferments did inside cells such as yeast.

In 1896, however, the German chemist Eduard Buchner (1860–1917) showed that it was possible to mash up yeast cells, rupturing their cell walls and freeing the protoplasm within. He left not a single intact cell and yet the fluid he obtained could do all the work of the intact cells. It became clear that anything that could work inside the cell could work outside the cell, too. The term *enzyme* became universal for any catalyst associated with living tissue.

As research continued, it turned out that virtually every chemical reaction that proceeded in living tissue was mediated by an enzyme—a different enzyme for each reaction.

The question arose, then, as to what the enzymes might be, chemically. It seemed logical to suppose them to be proteins, for only proteins had the kind of structure that could produce the many thousands of different but related molecules that were required for all the enzymes there seemed to be in all forms of life. The German chemist Richard Willstätter (1872–1942) showed during the 1920s, however, that solutions of enzymes that showed pronounced catalytic properties also proved to yield negative results to the most delicate protein tests known.

This was not really conclusive, since catalysts are active in such tiny concentrations that the enzymes might be proteins and yet be present in too small an amount to react to the tests. In 1926 an American biochemist, James Batcheller Sumner (1887–1955), working with preparations of an enzyme called *urease,* carefully concentrated the preparation, making it richer and richer in the enzyme, until he obtained tiny crystals. These, when dissolved in water, showed strong urease properties. The enzyme was, under such conditions, sufficiently concentrated so that, when tested, it proved to be unmistakably protein in nature.

Other enzymes were crystallized within the next few years and these proved also to be proteins. It quickly became apparent that all enzymes were proteins.

The importance of proteins could now be seen. It was the individual enzymes in each cell that controlled the various intermeshing chemical reactions within the cell. It was because one enzyme might be present and one absent, or because one was present in greater concentration and one in lesser, or because one was more efficient and one less, or one was kept under wraps and one was stimulated that cells existed with different properties and abilities.

That was why some cells were muscle cells and some nerve cells and some liver cells and so on. That was also why some were mouse liver cells and some were rat liver cells and some were mackerel liver cells and some were human liver cells.

That was also why one egg cell might develop into a grizzly bear and another into a dolphin. The egg cells looked the same, but the enzyme content varied. That was why one species looked different from another, and one individual within a species looked different from another.

Naturally, the enzyme patterns in cells of different individuals of a particular species resemble each other more closely than the enzyme patterns in different species. And within a species, the enzyme patterns in different members of a particular family resemble each other more closely than the enzyme patterns in unrelated individuals.

But what controls the nature of the enzymes in a particular organism? And what makes it certain that the enzymes in a child bear a particularly close resemblance to those in his parents?

By the 1930s it seemed quite clear that it must be the chromosomes that somehow controlled the nature of the enzymes. An offspring inherited a half-set of chromosomes from one parent and a half-set from the other, and so it resembled each parent—but not exactly.

How did the chromosomes determine what enzymes a new cell or a new organism were to have? The chromosomes were also protein; nucleoprotein to be exact. Biochemists, at first, did not place much emphasis on the nucleic-acid portion of the chromosome. It was, after all, not uncommon for proteins to do their work in association with nonprotein molecules.

The nonprotein molecules were, however, invariably far simpler in structure than the protein itself. The nonprotein molecule, called a *prosthetic group,* or a *coenzyme,* might have some subsidiary function, but it was always the protein molecule itself (or so it seemed) that possessed the capacity for enormous variation and made it possible to differentiate between organisms and between species.

At the start, nucleic acids, too, seemed to be much simpler than proteins. They, too, were polymeric molecules and were built up of relatively simple units called *nucleotides*. The nucleotides, to be sure, were more complex than the amino acids of proteins, but there were only four different nucleotides making up the nucleic acids. Even four different units is quite remarkable for a polymeric molecule, but how could it be compared with the twenty different amino acids making up proteins?

The different nucleotides have names, of course, but it is not important in this book to dwell on any terminology we can reasonably avoid. Biochemists usually refer to the different nucleotides by the initials of their names, and that will do for us. Each DNA molecule contains four different nucleotides: A, G, C, and T. Each RNA molecule contains four different nucleotides: A, G, C, and U. (T and U are very similar, but even a slight difference can be important in the chemistry of life.)

For quite a while, it was thought that each nucleic acid consisted of only four nucleotides altogether, one of each variety. That made nucleic acid molecules much smaller than protein molecules and reinforced the notion that it was the protein and not the nucleic acid that was the important component of chromosomes.

There was, to be sure, some evidence that was unsettling. Chromosomes in different cells might possess different amounts of protein, but they always had a fixed amount of nucleic acids. Sperm cells are very small, so that one might imagine they would have to get rid of all nonessentials—and in them the protein content was unusually small, but the nucleic acid content still remained fixed.

What's more, biochemists began to realize that the ordinary methods of isolating nucleic acid were too rough. By using such methods, they ended not with the molecules themselves, but with small shreds of them. Once gentler methods were used, it turned out that intact nucleic acid molecules were every bit as large as protein molecules; indeed, larger.

Nevertheless, it was difficult to turn away from proteins as the central molecules of life.

The answer came from bacteriology.

Bacteriologists were working with two strains of a bacterium that caused pneumonia. One strain had a smooth pellicle around each bacterial cell and was referred to as an *S-strain* (for "smooth"). The other lacked the pellicle and was the *R-strain* (for "rough"). Apparently, the S-strain possessed a chromosome fragment, or gene, that served to produce the pellicle, while the R-strain lacked it.

A British bacteriologist, Fred Griffith (1881–1941), who first worked with these strains, discovered in 1928 that if dead S-strain bacteria were mixed with live R-strain, the R-strain would develop pellicles. That made it seem that even when the S-strain bacteria were dead, the gene within them that produced pellicles could still do its work. The gene was referred to as the *transforming principle*.

A Canadian-American physician, Oswald Theodore Avery (1877–1955), worked on the S-strain bacteria, trying to isolate and purify the transforming principle and in 1944 finally managed to get an extract that contained no protein at all. It contained only DNA, and yet that solution of DNA served to convert R-strain to S-strain. This was the first indication that nucleic acid, and not protein, was the functioning part of a gene.

Since chromosomes doubled their number within the cell during cell division, each chromosome must have some system for forming an exact replica of itself *(replication)* so that the daughter cells would have the same genes as the mother cell. All the studies of proteins over the course of the previous century had never shown any of them to possess the power of replication. If it was DNA, and not proteins, that was the key component of genes and chromosomes, might it be that DNA *was* capable of replication?

Chemists began to study nucleic acid molecular structure in detail to see how this replication might take place. In 1948, for instance, the Austrian-American biochemist Erwin Chargaff (b. 1905) found that in DNA molecules A-nucleotides occurred in the same numbers as T-nucleotides, while G-nucleotides occurred in the same numbers as C-nucleotides.

In Great Britain, meanwhile, an English physical chemist, Rosalind Elsie Franklin (1920–1958), was taking X-ray diffraction photographs of crystals of DNA. From the manner in which the X-rays bounced off the molecule, it was possible to deduce its repetitive features.

An American biochemist, James Dewey Watson (b. 1928), saw Franklin's photographs. Using them, he and a British physicist, Francis H. C. Crick (b. 1916), worked out the structure of DNA in 1953. It consists of strings of nucleotides, each arranged in a helix (the shape of a bedspring, or a spiral staircase). The two helixes were intertwined (a *double helix*) in such a way that a T-nucleotide on one helix always fit an A-nucleotide on the other, and a C-nucleotide on one helix always fit a G-nucleotide on the other. (This explained Chargaff's observations.)

Each nucleotide was, in a sense, the negative of the other, so that you might call one a ( + )helix and the other a (-)helix. During cell division, the two helices untwine and each one serves as a model on which a new helix is formed, with A's and T's always attracting each other and G's and C's doing the same. The original ( + )helix forms another (-)helix on itself, while the original (-)helix forms another ( + )helix on itself. The end result is that in place of one double helix, you get two double helices. Each of the two daughter helices are exactly alike, and both are like the original. In this way, replication takes place.

Although, ideally, replication should produce generation after generation of DNA molecules that are all exactly alike, in actual fact there are numerous reasons for the introduction of slight errors. As a result, different DNA molecules are perpetually being produced. Most such molecules are useless, but every once in a while a useful one is produced. It is these replication errors that produce the slight changes called mutations, and mutations are an important factor in evolution.

DNA replication seems to explain, satisfactorily, the principles of heredity, and one can scarcely avoid assuming that the DNA molecules control the production of enzymes. But *how* do the DNA molecules do this? The DNA chains are made up of four different nucleotides; the enzyme chains are made up of twenty amino acids. How can four nucleotides produce twenty amino acids?

The puzzle arises only if one supposes that each nucleotide must match an amino acid. That won't work. However, what if you consider *groups* of nucleotides? Suppose you consider nucleotide "triplets," three adjacent nucleotides. Since the nucleotides can follow each other in any order, any one of the four can be in first place, any one of the four in second, and any one of the four in third. That allows 4 x 4 x 4, or 64 different triplets: AAA, AAG, AAC, AAT, AGA, and so on.

If each triplet is associated with a particular amino acid, then there are enough triplets to allow two or three to be assigned to each amino acid. The pattern along even a tiny portion of the DNA content of a chromosome is quite complex enough to produce the pattern of an enzyme. Each gene then is responsible for the production of an enzyme, and the enzyme content of a cell defines the properties and abilities of that cell. DNA replication ensures that the properties and abilities of a daughter cell are those of the parent cell, and that the properties and abilities of an offspring are those of their parents.

In the years since 1953, biochemists have worked out the *genetic code* by determining which nucleotide triplet stands for which amino acid.

To be sure, the DNA molecules are in the nucleus, while the ribosomes, which are the sites of enzyme manufacture, are in the cytoplasm. The information contained in the DNA must somehow get out to the cytoplasm.

This is done by transferring the DNA information to RNA, since RNA is in both nucleus and cytoplasm. A DNA helix can produce an RNA molecule replicating its structure. This *messenger-RNA* carries the DNA pattern to the ribosomes. There, numerous relatively small RNA molecules attach themselves to the messenger-RNA. The small RNA molecules come in a number of varieties, each of which has the capacity to fit itself to one particular triplet. The other end of the RNA molecule can fit itself to one particular amino acid. The various amino acids then combine on the ribosome and carry within themselves the DNA pattern as translated into amino acids. The small RNA molecules that transfer the nucleic acid information at one end of their structure to amino acid information at the other are called *transfer-RNA*.

It would seem then that if we're talking about the beginning of life, we could boil it down to the appearance, somehow, of a DNA molecule sufficiently complicated to be able to undergo replication. From that, everything else would follow.

But that's not so easy. DNA is an extraordinarily complex molecule, and to do its work, it needs the help of enzymes. This leads us into a kind of catch-22 situation. In order to have enzymes, you must first have DNA, but in order for DNA to do its work, you must first have enzymes.

In order to break out of that situation, there must be some simpler system out of which DNA arose, one which does not require enzymes to begin with. There are reasons for supposing that that simpler system involves the use of RNA.

For one thing, DNA exerts its influence through RNA, and it would seem that RNA does the actual work of enzyme synthesis while DNA is only the information-store. We can easily visualize a primitive situation in which RNA was both the information-store and the working mechanism.

This is not just a matter of imagination. The more complex viruses contain DNA, but the simpler ones, like tobacco mosaic virus, contain only RNA—no DNA at all.

One of the complexities of replication is that it requires a double helix, so that each of the two helixes can guide the formation of its partner. But is that an absolutely essential complexity? The American biophysicist Robert Louis Sinsheimer (b. 1920) discovered a strain of virus that contained DNA made up of a single helix, or *strand*, yet that DNA could replicate.

The method was simple enough. Imagine the single strand to be a ( + )helix. It could form a (-)helix, which could, in turn, from a ( + )helix. The replication is done in two steps rather than one and ends in a single new molecule, rather than in two. Single-strand DNA is a far less efficient replicator than double-strand DNA is, but it works just the same.

It might seem then that single-strand RNA is the original form of nucleic acid replicator. What's more, the shorter that single strand, the more rapid the replication and the simpler the process as a whole. Apparently, the replication of a single-strand RNA made up of less than a hundred nucleotides is a process so simple that it can proceed without the help of enzymes.

We might visualize the beginning of life as follows, then:

1. A very short single-strand RNA molecule that can replicate itself without enzymes and catalyze the formation of simple protein molecules.

2. The RNA molecule associates itself with some of the simple proteins it has formed, or some that have formed otherwise, and is in this way rendered more stable. The molecule can grow longer and replicate more efficiently.

3. The DNA molecule is formed, perhaps through an error in RNA replication. It is stabler than the RNA molecule, can exist in much longer chains (up to millions of nucleotides), and can store information more securely and replicate more efficiently and in a more error-free manner. The association with protein grows steadily more complex and useful.

4. These viruslike forms finally develop into simple prokaryotes, and from these come everything else.

This brings us to the next stage in the problem. How did the original single-strand RNA molecule come into being in the first place?

The question of the origin of life, if one omits the possibility of a supernatural creation, involves the passage from a substance that is definitely unliving to one that is, in however simple a fashion, alive.

In ancient times, this would not have been seen as a problem. Maggots appeared in rotting meat out of nowhere, for instance, and one could only assume that the rotting meat, clearly dead, was somehow converted into maggots, clearly alive. It was only when careful observation showed that the maggots formed only after flies had laid their eggs in the meat that this example of *spontaneous generation* was found to be not spontaneous at all.

During the nineteenth century, it came to seem more and more certain that all life originated from earlier life. In 1864, Pasteur showed this to be true even for microorganisms.

And yet life, at the very beginning, had no earlier life to start from. The boundary line separating nonlife from life must have been crossed.

Scientists, having decided that spontaneous generation simply did not take place, were reluctant to accept the necessity of supposing that it had taken place at some long-distant time in the past. In 1908 the Swedish chemist Svante August Arrhenius (1859–1927) tried to strike a compromise by supposing that life on Earth had originated when spores (living, but capable of very long periods of suspended animation) drifted across space for millions of years, perhaps until some landed on our planet and were brought back to active life by its gentle environment.

This is highly dramatic, but even if we imagine that Earth was seeded from another world, which, long, long before, had been seeded from still another world, we must still come back to some period when life began on *some* world through spontaneous generation. And since we must deal with spontaneous generation somewhere and at some time, we might as well see if we can deal with it here on Earth during the first billion years of our planet.

Why not? Even if spontaneous generation does not (or, possibly, *cannot)* take place on Earth now, conditions on the primordial Earth were so different that what seems a firm rule now may not have been so firm then. For instance, we now have an atmosphere rich in oxygen, but the primordial Earth had one in which oxygen was absent. That might well make an important difference.

Then, too, if we imagine life in the process of formation nowadays, this proto-life would serve as food for uncounted numbers of the myriad life forms that now exist. It would never last. On the primordial Earth, without life, any proto-life that developed would continue to develop without interference—at least without that kind of interference.

Even so, the problem of explaining the beginning of life is difficult. The original molecules present on Earth, in the sea and atmosphere, that are of the proper nature and are present in sufficient quantities to serve as precursors of life are small ones made up of two to five atoms each. The simplest form of protolife we can imagine—a single-strand RNA molecule consisting of nearly a hundred nucleotides—would be made up of perhaps 3,700 atoms. Clearly, we are expecting life to begin by the building up of very small molecules into quite large ones.

The natural tendency, however, is for large molecules, left to themselves, to break up into small molecules. There is virtually no tendency for small molecules, if left to themselves, to build up into large ones. This is equivalent to saying that balls will readily roll down an incline but are not at all likely ever to roll *up* one.

Yet we need not imagine matters being left entirely to themselves. A ball may not roll up an incline of itself, but it can be *pushed* up an incline. What won't happen spontaneously may well happen if energy is supplied. In the same way, small molecules may build up into large ones if energy is supplied.

In the primordial Earth, there were energy sources—volcanic heat, lightning, and most of all, sunshine. Nowadays, some of the oxygen in the air forms ozone (an energetic form of oxygen with three atoms per molecule $[O_3]$, rather than $[O_2]$ as in ordinary oxygen). The ozone accumulates in the upper atmosphere and blocks the Sun's ultraviolet light. On the primordial Earth, with no oxygen in the atmosphere, there was no ozone layer, and the Sun's energetic ultraviolet would reach the Earth's surface undiluted.

The first person to consider the possibilities carefully was a Soviet biochemist, Alexander Ivanovich Oparin (1894–1980), who in 1936 published a book on the subject, entitled *The Origin of Life on Earth*. He considered the atmosphere on the primordial Earth to have been a mixture of methane and ammonia and the energy source to have been sunlight.

In 1954 a chemistry student, Stanley Lloyd Miller (b. 1930), working for the American chemist Harold Clayton Urey (1893–1981), tried to bolster speculation by experiment. He began with a mixture of water, ammonia, methane, and hydrogen that he made sure was sterile and had no life of any kind in it. He then circulated it past an electric discharge that would serve as an energy source. At the end of a week, he analyzed his solution and found that some of its small molecules had been built up to larger ones. Among these larger molecules were glycine and alanine, the two simplest of the twenty amino acids commonly found in proteins.

Others followed, using different mixes of what may have existed in the primordial sea and air, and using other energy sources. The results were much the same.

One of the products of such experiments was hydrogen cyanide (HCN). The Spanish-American biochemist Juan Oro (b. 1923) added hydrogen cyanide to his starting mixture in 1961. He obtained a richer mix of amino acids. He also obtained adenine, which is an important constituent of one of the nucleotides in nucleic acids. In 1962 Oro added formaldehyde (HCHO), another early product of these experiments, to his mixture and obtained a variety of sugars, including ribose, a component of RNA nucleotides, and deoxyribose, a component of DNA nucleotides.

Nor do these results take place only in experiments under the guidance of human beings, experiments that might, therefore, be unconsciously weighted in favor of life.

Most meteorites, for instance, are either metallic in

nature, or rocky, and neither type shows any trace of organic material. There is, however, a small percentage of meteors that are *carbonaceous chondrites* and that contain small quantities of water and of carbon-containing compounds. The Sri Lankese–American biochemist Cyril Ponnamperuma (b. 1923) has analyzed some of these and found traces of five of the amino acids that make up proteins.

Then, too, astronomers have been studying the radio waves emitted by vast clouds of dust and gas in interstellar space. From the nature of these radio waves, it is possible to tell what molecules have formed in these clouds. At first only two-atom combinations were found, but then, as radio telescopes grew larger and more efficient, other molecules were detected—water, ammonia, formaldehyde, methyl alcohol, and so on. If we could examine these clouds at close range, no one would be too surprised if we located amino acids or nucleotides.

This means that there is the possibility that the primordial Earth was given a "leg up," so to speak, with some simple compounds important to life being brought here by meteors or comets, or settling out into the atmosphere from surrounding dust.

However, no one has yet to get past the rather middle-sized compounds on the way to life. No experiments have even approached the compounds that would be required for even the most primitive form of life.

There are some recent suggestions that the reason is because life didn't form in a direct line from simple compounds to single-strand RNA capable of replication. One suggestion that has aroused some interest recently is that the true starting point is with some system that is capable of replication in ways far simpler than that of nucleic acids.

Inorganic crystals might conceivably fill the bill. Perfect crystals have orderly arrangements of atoms and are uninteresting. Real crystals, however, are never perfect, but always have defects in them, misalignments of atoms. These defects can propagate themselves in ways that amount to replication and can undergo changes akin to mutation. This does not in itself represent life or even a legitimate pathway to life, but it might offer a kind of model for something more suitable.

The British chemist A. G. Cairns-Smith proposes that clay might be the original replicating system. It is a common substance that readily forms crystals. Some organic substances can speed the formation of clay crystals and can attach themselves to the clay, forming clay/organic replicating systems. Those organic compounds that best fit the clay are "selected" so that the organic portion of the system slowly becomes more adept at replicating and begins to be the predominant part of the system. Eventually the organic portion can get by on its own and the clay is pushed to one side, so to speak, having served as a scaffolding that is no longer needed.

Suppose, then, that we start with the formation of Earth, 4.5 billion years ago. We can allow the first few hundred million years to pass while the Earth settles down to more or less its present state. It cools down and squeezes out an ocean and an atmosphere. The surrounding hydrogen is swept away by the solar wind, and the rain of meteors out of which the Earth was formed dwindles and virtually ceases.

Then, perhaps 4,000 million years ago, the Earth is reasonably quiet and the period of "chemical evolution" begins. Whether reasonably complex organic molecules developed directly from the small molecules of the air and ocean, or did so by way of clay or in any other fashion, the ocean was probably teeming with organic molecules by (perhaps) 3,800 million years ago. The ocean of the time is sometimes referred to as an "organic soup."

The primordial viruslike molecules (which we might call virusoids, though the name is not used by scientists to my knowledge) may have developed by then. These catalyzed the breakdown of the organic substances in the "soup," producing energy that made it possible to convert some of the surrounding compounds into more virusoids. The virusoid population grew and the organic soup, which served as food, tended to thin out.

Eventually, a balance might have been reached in which just enough virusoids existed so that the amount of food required to keep them alive was equal to that built up by the Sun's ultraviolet light. However, if the Earth's air was completely Atmosphere II by the time the virusoids existed, then water photolysis in the upper atmosphere was producing some oxygen and, therefore, some ozone. The ultraviolet light reaching Earth's surface would diminish, and if ultraviolet were an important energy source for the continuing production of organic matter in the ocean, the food supply would diminish.

The competition for food would sharpen, and those virusoids who could somehow amass a food reserve would win out. One way to do this would be to have a virusoid molecule with a membrane that would allow food molecules to be engulfed, but would not allow the molecules to diffuse outward again. In this way, a food supply would be accumulated within the membrane boundary, one that could be made use of at leisure. In short the virusoids would have to become cells.

The formation of cells is, perhaps, not a major problem. Beginning in 1958, the American biochemist Sidney Walter Fox (b. 1912) experimented with the effect of considerable heat on amino acids (heat such as might be expected on exposed rocks of a primordial volcanic earth, rocks that might be doused, periodically, with warm rain). He found that the amino acids joined to form a proteinlike polymer to which Fox gave the name of *proteinoid*. Dissolved in water, the proteinoids form tiny *microspheres,* bounded by membranes, and these microspheres displayed some of the properties we associate with cells.

It might be, then, that in the course of time, primordial virusoids combined with primordial microspheres to form the first very simple and fumbling prokaryotes soon after 3,500 million years ago.

Even if prokaryotes can store food, they still depend ultimately on the food supply in the oceans as built up by energetic ultraviolet light. If ultraviolet light is decreasing, then food is decreasing, and accumulating food stores only puts off the evil day of starvation. Therefore, any prokaryote that (by accidental mutation) takes a step toward being able to use the lesser energy of ordinary visible sunlight to manufacture larger molecules out of smaller ones has an advantage as far as survival is concerned. After all, visible light can, and does, get past any ozone barrier without trouble. If it can be used as an energy source it offers an unlimited source of food.

By 3,000 million years ago, or soon after, the cyanobacteria, the first organisms capable of photosynthesis were in existence. They could build up their own food from small molecules and did *not* depend on the ocean soup. Neither did the older bacterial prokaryotes, provided they developed methods for feeding on the cyanobacteria and using *their* food stores.

Photosynthesis, however, meant the consumption of carbon dioxide and the production of oxygen at a much greater rate than was possible merely by photolysis. The carbon dioxide of the atmosphere began to dwindle in amount while the oxygen content began to increase.

The presence of oxygen in the atmosphere hastened the demise of the ocean soup, since oxygen combined with organic molecules to form carbon dioxide and water. That meant that only cyanobacteria and those organisms that fed on them could survive in quantity. What's more, oxygen was dangerous even to cells unless enzymes were developed that could direct the combination of oxygen with organic molecules in a smooth and orderly fashion. Otherwise, oxygen would combine with cellular compounds at random and kill the cell.

Of course, even to this day, there survive some bacteria that are unable to make use of oxygen and to whom oxygen is, in fact, poisonous. They are *anaerobic bacteria* (Greek for "no air"). They exist only in the nooks and crannies of the environment but are by no means unimportant. There are anaerobic bacteria that can cause botulism, tetanus, and gas gangrene, all of them deadly diseases.

There are also bacteria that can obtain their energy from chemical reactions not involving photosynthesis *(chemosynthetic bacteria)*. Recently, such bacteria were found living in certain parts of the sea bottom where hot water, rich in chemicals, issued from vents. These bacteria supported an array of more complex life, all of which did not depend on the energy of sunlight and could live on even if all life on Earth's surface disappeared. They, too, however, only inhabit the nooks and crannies of the environment.

The process of oxygenation of Earth's atmosphere may have proceeded for a period of over 2,000 million years before virtually all the carbon dioxide was gone and the process came to a halt. The process was very slow at first, 1,400 million years ago, when eukaryotic cells formed; some (the algae) were photosynthetic, and much more efficiently so than the cyanobacteria. The rate of oxygenation speeded up and was essentially complete about 650 million years ago.

The direct use of oxygen to combine with organic molecules (thanks to the existence of appropriate enzymes) produced about twenty times as much energy for a given quantity of molecules as did the older processes of molecular breakdown not involving oxygen.

This meant that as the oxygen content of the air increased, life forms had a larger and larger supply of energy for use on what we might call luxuries. Life forms were able to devote a certain amount of energy to the development of hard parts for protection, for more efficient predation, for the attachment of stronger muscles, and so on, and that is why fossilization began so suddenly with the coming of the Cambrian period, 600 million years ago.

Yet the question of beginnings, even of life itself, cannot be confined to Earth, for there is much more to the Universe than our own planet. Suppose, for instance, we consider the Moon. How did it begin?

# 22

## MOON

The biblical tale of creation concerns itself primarily with the Earth and human beings. The remainder of the Universe is mentioned only in connection with its service to Earth and humanity and is quickly dismissed. Thus, on the fourth day of creation, the Bible says in Genesis 1:14-16,

> And God said, Let there be lights in the firmament of the heaven to divide the day from the night; and let them be for signs, and for seasons, and for days, and years: and let them be for lights in the firmament of the heaven to give light upon the earth: and it was so. And God made two great lights; the greater light to rule the day, and the lesser light to rule the night: he made the stars also.

The Moon was "the lesser light" and, until a few centuries ago, the feeling was undoubtedly general among human beings that it was merely a lamp hung in the sky for the convenience of humanity. It did not seem very far away, and it did not seem very large. The splotches visible on its surface were interpreted in various ways by various cultures. To us Westerners, it seemed "the man in the Moon," and the man was almost as large as the Moon—or, rather, the Moon was almost as small as a man.

Yet as long ago as 150 B.C., the Greek astronomer Hipparchus (190–120 B.C.) had worked out the distance to the Moon by trigonometric methods and had found it was sixty times the Earth's radius (that is, the distance from the Earth's center to its surface).

The Greek scientist Eratosthenes (276–196 B.C.) had already shown the Earth's circumference to be about 25,000 miles. The modern figure is 24,906 miles (40,075 kilometers). That meant Earth's radius, to use modern figures, is 3,964 miles (6,378 kilometers) and the distance to the Moon is 238,900 miles (384,400 kilometers). For the Moon to seem as large in the sky as it does at that distance, it must be 2,160 miles (3,476 kilometers) in diameter.

In other words, the Moon is a little more than a quarter the diameter of the Earth. It is not merely a lamp in the sky. It is a sizable world, and this was known to Hipparchus twenty-two centuries ago.

This must have seemed nothing but rarefied philosophical speculation to the ordinary person (if he heard about it at all). In 1609, however, Galileo turned his telescope on the Moon and saw mountains, craters, and what looked like seas. Thereafter, there was no question that the Moon was a world.

Once Newton had worked out the law of universal gravitation in 1687, it was possible for him to demonstrate that the ocean tides are caused by the gravitational pull of the Moon, which decreases in intensity with distance.

The Moon's gravitational pull is, therefore, a little stronger on the side of the Earth toward itself than it is upon the side away from itself. This results in a stretching of the Earth along the line that connects its center with the Moon's center, and the creation of two bulges on either side, the water being stretched more than the rocky crust is. (The gravitation of the Sun also contributes to the tides.)

As the Earth turns, so that different portions of its

surface pass, progressively, through the bulges of water, the water scrapes against the shallower sea-bottoms and converts some of the turning energy of the Earth into heat, through friction. This slows the Earth's rotation to a very slight degree, lengthening the day by 1 second in the course of 62,500 years.

This is not much, but rotational momentum cannot be destroyed; it can only be shifted elsewhere. If the Earth's rotation slows, the Moon's turning motion about the Earth must be increased. One way of doing that is to have it move farther off so that it must turn through a longer orbit. It follows that the Moon's tidal effect is very slowly forcing it farther from the Earth.

That point was used to make the first scientific attempt at reasoning out the beginning of the Moon. As I mentioned earlier in the book, Buffon had speculated that the Moon had been ripped out of the Earth early in its history, but it was *just* a speculation. He had no clear line of reasoning, no evidence, to justify the statement.

The English astronomer George Howard Darwin (1845–1912), the second son of the biologist Charles Darwin, tried in 1879 to use the tidal effect to justify Buffon's speculation of a century earlier.

Darwin pointed out that if one looked into the past, the Moon must then have been closer to the Earth, and the Earth would have been rotating more rapidly. In fact, if one looked back into the past far enough, the Moon was close enough to Earth to be part of it.

In other words, Darwin maintained that the Moon and Earth formed a single body in the days when Earth was first formed. The Earth was then spinning so quickly, however, that the centrifugal effect produced a huge equatorial bulge. Part of the Earth's equatorial region bulged farther and farther away, forming a kind of dumbbell shape, with one side much larger than the other. Finally, the smaller portion, about one-eightieth the mass of the whole, broke away, forming the Moon. Thanks to tidal action, the Moon moved farther and farther away, and the Earth's rotation period has slowed and slowed ever since.

(The Moon's own rotation slowed even more rapidly than Earth's did because the larger Earth exerts a larger tidal influence on it than it does on us. What's more, the Moon, being smaller, has less rotational momentum, so that it bleeds away more quickly. In any case, the Moon's rotation has now slowed to the point where it faces one side permanently toward the Earth.)

This picture of the origin of the Moon is a very attractive one in some ways. If it is true, the Moon would have been formed from the upper layers of the Earth, which are distinctly lower in density than the Earth as a whole is. (That is because the center of the Earth seems to contain a huge nickel-iron core that increases the general density of the planet but was not affected by the splitaway.) And, to be sure, the Moon is only three-fifths as dense as the Earth is, about as dense as the rocky "mantle" of the Earth that lies outside the nickel-iron core. The Moon has no nickel-iron core of its own.

Again, the Moon is just about as wide across as the Pacific ocean is, so that one can imagine that it was pulled out from where the Pacific ocean is now located, leaving a vast hollow to be filled with water. The scar of that involuntary surgery might still show in the band of volcanoes and earthquakes that rim the Pacific today.

However, Darwin's theory did not hold up. We know the amount of spin in the Earth-Moon system. We know exactly how much spin there is in the Earth's rotation about its axis, the Moon's rotation about its axis, and the Earth-Moon revolution about their mutual center of gravity. If all this momentum of spin were concentrated into a single body that had the mass of Earth and Moon put together, and that was spinning about its axis, that body would still not have enough spin to split in two. Therefore Darwin's picture had to be dismissed.

What's more, the Pacific's shape today, and the earthquakes and volcanoes that rim it, have been satisfactorily explained by plate tectonics and have nothing to do with the Moon.

The alternative is that the Moon was formed separately

from Earth to begin with. But if that is so, where might it have been formed? If it had been formed close to the Earth to begin with, it should be revolving in nearly the plane of Earth's equator, but it isn't. It is revolving, instead, in nearly the plane of Earth's orbit about the Sun, as though the Moon had once been an independent planet and had been captured.

However, if the capture suggestion is true, it would represent a most unusual situation, for it would be very hard for the Earth to capture a body the size of the Moon. Astronomers have not yet figured out a good set of circumstances for that to happen. What's more, if it had been captured, it is likely that it would have a more elliptical orbit than it has now.

On the other hand, if the Moon could not have been captured and if it were formed in the neighborhood of Earth, it would have to be formed out of the same materials as the Earth was. Why, then, does it not have a nickel-iron core? Astronomers have not yet figured out a good way of explaining why all that iron and nickel should have been grabbed by the Earth and virtually none by the Moon.

Beginning in 1969, astronauts have been landing on the Moon and have been bringing back rocks from this satellite of ours. The hope was that a close study of the rocks might settle the matter. It is clear from those rocks that the Moon is as old as the Earth, but where it might have been located when it was formed remains an open question, despite all that the rocks can tell us.

Some astronomers, in disgust, have said that since all three possibilities for the origin seem to be unlikely, the only logical conclusion is that the Moon really doesn't exist.

It isn't quite as bad as that, however. What was needed was a fourth possibility. As early as 1974, an American astronomer, William K. Hartmann, suggested one. He said that perhaps a large body had struck the Earth a glancing blow early in its history, and that the Moon had originated in that fashion.

The suggestion was largely ignored then, but by 1984 it had been backed by computer simulations and it began to look better and better. Now it has become quite popular.

The suggestion is that the interloper was about as big as Mars, or even a little bigger perhaps, and had one-seventh the mass of the Earth. It struck the Earth soon after our planet had assumed its present state and before any life had appeared on it. (If life had existed, the collision would have wiped it out.) It must have happened over 4,000 million years ago.

The blow of the interloper would have vaporized much of the surface layers of both worlds and sent them shooting out into space. Much of what was left of the interloper remained fused to Earth and the two finally settled down into a single body. The material that vaporized soon cooled and solidified into bodies of different size that gradually coalesced and formed the Moon.

This would account for the Moon's plane of revolution about the Earth not being in the plane of Earth's equator, for that plane would depend on the exact angle at which the intruder hit. The new suggestion would account for the Moon's absence of a nickel-iron core because only the outer layers of the two worlds vaporized and formed the Moon. The cores remained relatively untouched. It would also account for the fact that the Moon is short on substances that are easily vaporized because it would have formed out of hot matter, and material that was too easily vaporized would not readily solidify and had time to vanish into the far reaches of space.

In short, just about all the puzzles of the Moon's origin that the first three alternatives couldn't solve are indeed solved by the new collision hypothesis. The hypothesis may not survive, but it looks good at the moment.

One question, though, is this. Where did the interloper come from?

To answer it, we have to realize that the Earth is not alone in space. It is part of a large family of objects that includes the Sun and the various planets and other bodies that circle the Sun—bodies as large as giant Jupiter and as small as a microscopic dust particle. The whole family of objects is called the *solar system* (from the Latin word "sol" meaning "sun").

Let us ask about the beginnings of the Solar system and see if that might help us explain where the interloper came from.

# 23

## SOLAR SYSTEM

In ancient and medieval times, it was taken for granted that the Earth was the center of the Universe, for the very good reason that it *seemed* to be. Seven bodies, *or planets,* were thought to circle the Earth at progressively greater distances—the Moon, Mercury, Venus, the Sun, Mars, Jupiter, and Saturn. Beyond that was the black sphere of the sky, on which the glowing sparks of the stars seemed to be fixed.

It was not until 1543 that this view was fundamentally changed. In that year, the Polish astronomer Nicolas Copernicus (1473–1543) published a book pointing out that the mathematics of calculating planetary movements would be simplified if one assumed that all the planets (including the Earth and its attendant Moon) were revolving about the Sun. Some ancient Greek astronomers had suggested this, but Copernicus was the first to develop the notion mathematically.

It took over half a century, however, to overcome the ancient habits of thought, and even as late as 1633 Galileo was forced by the Inquisition to deny publicly that the

Earth moved. It moved, just the same (as Galileo is popularly supposed to have muttered under his breath), and that action was the last gasp of the old Earth-centered belief—at least among the scientifically literate.

In 1609 the German astronomer Johann Kepler (1571–1630) had demonstrated that the orbits followed by the planets around the Sun were not circles, as had been thought, but ellipses, with the Sun at one focus. In this way, the nature of the planetary system was established in the form that is accepted to this day.

The Sun is at the center of the planetary system, then, and we now know it to be a huge body, 332,800 times as massive as the Earth, and 743 times as massive as all the objects, from planets to dust, that circle it. It dominates everything else to such an extent that it is only reasonable to speak of the entire collection of bodies as the solar system.

The solar system exhibits certain regularities. The planets all circle the Sun in the same direction and all do so, more or less, in the same plane, that of the solar equator. Almost all the planets, and the Sun, too, rotate about their axes in the same direction as they revolve about the Sun. The satellites, for the most part, also revolve about their planets in this same direction, and usually do so in or near the equatorial plane of the planet they circle.

This sort of thing tends to make scientists suppose that the solar system was not formed at different times and under different conditions, since that could scarcely impose this apparent uniformity of structure. Rather, the solar system must have been formed by some one single action that produced all the bodies either at once, or at regular intervals under similar conditions.

In 1745, Buffon, who was the first to suggest a considerable age for the Earth, also suggested a method whereby the solar system might have been formed. He thought that a massive body must have struck the Sun many years ago, and that solar debris was flung far out into space as a result. The debris cooled and formed the planets.

By that notion, all the planets were formed at the same

time, while the Sun itself was older than the planets, possibly much older.

This is not actually such a bad idea. It is very similar to the current idea, described at the end of the previous chapter, that is being advanced to explain the formation of the Moon. However, Buffon's suggestion was not taken up by astronomers, for it was only a speculation. Buffon had no evidence to offer in its favor.

In 1755 the German philosopher Immanuel Kant suggested something altogether different. Building, perhaps, on a notion casually put forth by Isaac Newton about seventy years earlier, Kant supposed that the solar system had started as a vast cloud of dust and gas that slowly came together to form a compact body—the Sun.

The particles of matter, moving inward under the influence of the gravitational field of the cloud, would gain energy of motion from that field. (Energy of motion may be called *kinetic energy*, from a Greek word meaning "motion".) When the motion was brought to a halt with the formation of the Sun, kinetic energy was converted to heat, and it was this heat that has caused the Sun to glow ever since.

That suggestion also failed to stir much interest. Again, there was no evidence, so that it was merely a speculation. In 1798, however, the French astronomer Pierre Simon de Laplace (1749–1827) advanced the same idea at the end of a book on astronomy intended for the general public. Laplace may not have known of Kant's earlier suggestion, and, in any case, he went into greater detail.

Laplace suggested that the original cloud of dust and gas was spinning. As it condensed, it would spin faster and faster, according to the well-known *law of conservation of angular momentum*. Eventually, it would spin so fast that it would flatten out into a lens-shaped body and the material at the far end of the lens would drift away under the influence of a centrifugal effect. The material that drifted away would cool and condense into a planet.

The loss of the planetary matter would carry off some of the spin and the main mass of the cloud would slow its rotation. As the cloud condensed further, the spin would quicken again until another shell was cast off, and so on. In this way, a whole series of planets would be formed, each rotating on its axis and revolving around the Sun.

Laplace's suggestion seemed to take care of all the details. He could even point to an example of what he was talking about.

In the constellation of Andromeda, there is a small cloudy patch that had first been described, in 1611, by the German astronomer Simon Marius (1573–1624). It was called the *Andromeda nebula* (from a Latin word meaning "cloud"). Laplace suggested that the Andromeda nebula was a cloud of dust and gas that was slowly condensing into a planetary system like our own. As a consequence, his description of the formation of the solar system came to be known as the "nebular hypothesis."

By the nebular hypothesis, the outermost planet is the oldest and the planets grow younger as one moves toward the Sun. Thus, Mars would be older than Earth, which would, in turn, be older than Venus. The Sun would be the youngest of all the bodies in the solar system.

The nebular hypothesis caught the imagination of astronomers and of the general public, so that, for nearly a century, it was accepted as the probable way in which the solar system was formed.

A number of minor matters seemed to fit the nebular hypothesis and to strengthen it. The planets themselves, in forming, might cast off smaller rings of their own to form the satellites.

Saturn actually has a set of rings circling it, which are closer to the planet than any of its visible satellites are. In 1859 the Scottish mathematician James Clerk Maxwell (1831–1879) showed that those rings were not solid, but consisted of small particles. This seemed to be an example of what Laplace was talking about.

When the small bodies of the asteroid belt were discovered, beginning in 1801, that also seemed to be a case of a ring of matter that had never had the chance to coalesce—

perhaps because of the disturbing effects of nearby Jupiter's gravitational field.

Helmholtz's theory of the Sun gaining its energy by a slow shrinkage seemed also to fit in with Laplace's hypothesis.

But then came the matter of spin, or angular momentum. George Darwin's theory of the Moon splitting off from a rapidly spinning Earth fell through because there was not enough angular momentum in the Earth-Moon system to allow that to happen. In the case of the nebular hypothesis there was an opposite problem. There was too much angular momentum in part of the solar system.

The planets make up only a little over 1 percent of the mass of the solar system, yet the angular momentum of the planets is 98 percent of that of the entire system. Jupiter has 60 percent of the total all by itself. The Sun possesses only 2 percent of the angular momentum of the solar system, so that Jupiter has thirty times as much angular momentum as the much larger Sun does.

How is it possible for so much angular momentum to be concentrated in the planets? When the spinning cloud of dust and gas began condensing in accordance with the nebular hypothesis, it had to possess all the angular momentum of the system. Some of it was bled off with each ring of matter given off, but there was no way of figuring out how 98 percent of the total could be crammed into those rings of matter.

This problem seemed insoluble and astronomers were forced to abandon the nebular hypothesis by the end of the nineteenth century. Yet the solar system must have had a beginning. If not the nebular hypothesis, something else would have to be worked out. Astronomers' attention therefore returned to Buffon's suggestion of formation by collision rather than by condensation.

In 1900 two American scientists, Thomas Chrowder Chamberlin (1843–1928) and Forest Ray Moulton (1872–1952), worked out the consequences of another star passing quite near the Sun (an actual collision, they thought, might not be necessary). The gravitational pull between them would drag out a mass of material that would stretch between the two stars as they pulled away from each other.

The hot matter pulled out of the Sun and the other star would condense into relatively small objects called *planetesimals*. These would move about the Sun in a chaotic variety of orbits, and there would be frequent collisions. On the whole, as a result of those collisions, the larger bits would grow at the expense of the smaller ones until finally there would be the planets we now know. The Chamberlin-Moulton notion is therefore called the "planetesimal hypothesis."

As to the matter of angular momentum, the English astronomers James Hopwood Jeans (1877–1946) and Harold Jeffreys (b. 1891) pointed out that as the two stars separated, the gravitational fields would give the mass of pulled-out matter a sidewise yank. This would pile angular momentum into them at the expense of the two stars. This gave the planetesimal hypothesis a big boost.

The planetesimal hypothesis goes back to Buffon's notion that the Sun existed before—perhaps long before—the planets were formed, and nothing was said about when or how the Sun formed.

During the early 1900s, the planetesimal hypothesis was accepted by many astronomers. In the 1920s, however, the English astronomer Arthur Stanley Eddington (1882–1944) showed that the interior of the Sun was much hotter than anyone had expected it to be. Its temperature at the center would be in the millions of degrees. Only with such temperatures in the interior could the Sun keep from condensing into a tiny body under the pull of its own gravity. (Such central temperatures turned out to be necessary when, ten years later, it was argued that the Sun's energy arose from nuclear fusion.)

This meant that the material pulled out of the stars at a close approach would have been much hotter than the upholders of the planetesimal hypothesis had counted upon. In 1939 the American astronomer Lyman Spitzer, Jr., (b. 1914) showed that the matter from the stars would be so hot that it would simply expand into the vacuum before it had a chance to condense. There would be no planetesimals and no planets.

There were other problems, too, with working out mechanisms to make sure the planets had enough angular momentum and were able to take up orbits sufficiently far from the Sun. The hypothesis kept being modified but nothing made it work, and by 1940 it was dead.

But then, in 1944, the German astronomer Carl Friedrich von Weizsacker (b. 1912) returned to the nebular hypothesis, with new mathematical tools.

He pictured a cloud condensing, just as Laplace had pictured it, but instead of giving off rings of gas, it condensed more rapidly, leaving a large disc of gas and dust around it. Within this disc there were turbulent eddies and sub-eddies.

These whirling eddies would carry material into collisions in their regions of intersection, forming planetesimals that would grow larger and larger with continuing collisions until the planets were formed. The mathematical treatment showed how the planets would form at increasing distances from each other as the eddies grew progressively larger with greater distance from the Sun.

Weizsacker's hypothesis grew quickly popular. By it, the Sun and all the planets would seem to have formed at roughly the same time. We can therefore conclude that the entire solar system is about 4,550 million years old, or a little older if we count the planetesimal period before it. This is borne out by the ages determined for various meteorites and for the oldest rocks obtained from the Moon.

That still leaves the question of angular momentum. The Swedish astronomer Hannes Alfven (b. 1908) took into account the magnetic field of the Sun, which had till then been neglected by those working out methods of formation of the solar system. As the young Sun whirled rapidly, its magnetic field twisted with it and acted as a brake, slowing it up. This meant that angular momentum would pass from the Sun to the planets, forcing the planetary orbits farther from the Sun.

This new version of the nebular hypothesis is now accepted by astronomers generally and seems to leave no major problems unanswered. Just as in the hypothesis of Chamberlin and Moulton, the planets formed from planetesimals that were gradually swept up out of the planetary orbits. Even when the planets were roughly their present size, there remained the last few planetesimals to sweep up. The final collisions left their marks behind in the form of craters.

We are familiar with such craters. The Moon's craters have been known since Galileo first looked at the Moon with his telescope. They were formed, for the most part, 4,000 million years ago, when planetesimals were still common, but some were formed more recently, for even now collisions are not unknown. In this age of planetary probes, we have also found craters on other airless or nearly airless worlds, such as Mercury, Mars, and various satellites.

Planets with atmospheres are not so rich in craters because craters there tend to be eroded by wind. On Earth, there is also the effect of water and of life, so that our planet has almost no craters produced by collisions. In Arizona, there is a crater half a mile across that may have been produced by the fall of a rather large meteor fifty thousand years ago. There are traces of older craters, too, that are nearly eroded away. About 65 million years ago, a particularly bad collision may have been the cause of the death of the dinosaurs and many other types of life-forms at the end of the Cretaceous.

Back before 4,000 million years ago, when the last of the large planetesimals were sorting themselves out to see which would survive as planets, one that was roughly the size of Mars may have struck the Earth in such a way as to form the Moon. *That* is the answer to the question as to where that interloper came from. It was one of the last survivors of the age of the planetesimals and it might have made it to independent planethood, as Mars did, if it had not had the misfortune of colliding with the still larger Earth.

One important difference between the nebular hypothesis, in any form, and the planetesimal hypothesis, is this. If the nebular hypothesis is true and if a planetary system forms by the condensation of an original cloud of dust and gas, then perhaps all stars form like this and all stars may have planets of one sort or another. On the other hand, if the planetesimal hypothesis is true and if a planetary system forms by a close passage of two stars, then, considering how far apart stars are and how slowly they move compared to the distances between them, such passages happen very, very rarely. In that case, the solar system is very much an exception and very, very few stars may be expected to have planets.

In the last few years, a satellite equipped to detect infrared light has detected such light in the near neighborhood of some stars. The infrared-light radiation is the mark of relatively cool matter, so it would seem that those stars are surrounded by cool matter. Close analysis makes it seem that stars such as Vega and Beta Pictoris are surrounded by a region of planetesimals within which planets may be forming or have already formed. This is an important strengthening factor for the current picture of solar-system formation.

This is a reminder, by the way, that the Sun is only one of many, many stars. Granted that every star may have formed much as the Sun did, that means that before any stars existed, the whole Universe must have consisted of a vast quantity of dust and gas. How did this come into existence?

What, in other words, were the beginnings of the entire Universe? That is our last question.

# 24

## UNIVERSE

Since it seems now to be the case that the whole solar system had its beginnings at the same time, some 4,550 million years ago, is it possible that at that time all the other stars had their beginnings, too?

The answer to that is no. Let's reason it out.

Over the years, astronomers have gradually learned a great deal about the stars. It is not necessary in this book to go into great detail about all these discoveries, but let's refer to those which play a part in determining how and when the Universe began.

Until modern times, it seemed that the stars were simply luminous objects attached to the solid sphere of the sky. In the 1600s, the nature of the solar system had been worked out and the distances separating the Sun and planets from each other were roughly known. It was plain that the solar system, as far out as Saturn (which was the farthest planet known prior to 1781), was at least 1,800 million miles (2,800 million kilometers) across, but it still remained possible that the sky was a sphere just a little larger than that in diameter, with the stars attached to it.

The turning point came in 1718 when Edmund Halley noted that three of the brightest stars had changed their positions relative to the remaining stars. That made it seem as though the stars were not fixed to some solid sphere, but moved independently like a swarm of bees. They were so far away that the motions were barely noticeable, and, naturally, the nearest (and therefore the brightest) would move more noticeably than the others.

But, then, if the stars were very far away, just how far away might that be? Actually, Halley made an estimate of that. He supposed that Sirius was actually an object as bright as our Sun. How far away would it have to be to appear no brighter in the sky than it does? Halley calculated it would have to be about 12 trillion miles (32 trillion kilometers) away, where a trillion is equal to a million million. Since light travels 5.88 trillion miles (9.46 trillion kilometers) in one year, that distance is called a *light-year*. Halley was saying that Sirius was about 2 light-years away. (Actually, Sirius is considerably brighter than the Sun, so that it must be over four times as far away as that to appear the mere spark of light it does.)

Is there anything we can do that would be better than a guess? Yes, we could measure the tiny shift of the nearer stars relative to the farther ones as the Earth shifts its own position from one side of the Sun to the other. This shift of the apparent position of an object with the shift of the observer is called the *parallax* of the object. The larger the parallax, the smaller the distance of the star. It is easy to make the calculation once the parallax is observed, but that observation is difficult. The telescopes of Halley's day were not good enough.

The first to report a stellar parallax was the German astronomer Friedrich Wilhelm Bessel (1784–1846), who announced in 1838 the parallax of a star named 61 Cygni. From that, he calculated its distance. The best figure we now have of that star's distance is 11.2 light-years, so that it takes 11.2 years for light to travel from 61 Cygni to us.

As it happens, 61 Cygni is not the nearest star. In 1839 the Scottish astronomer Thomas Henderson (1798–1844) reported Alpha Centauri to be 4.3 light years away. Alpha Centauri is actually a three-star system, and one of the three, Proxima Centauri, is, to a slight degree, the closest star to ourselves, so far as we know.

Nowadays, of course, we know the distance of stars much farther off than Alpha Centauri or 61 Cygni.

It turns out that the nearer stars are not invariably brighter than more distant stars. This would be so if all stars were equally luminous (that is, if they all gave off the same amount of light), but they are not. A very luminous star would be bright even from a great distance, while a star of low luminosity would be dim even if it were fairly close.

Thus, Proxima Centauri, although the closest star, is so dim that it cannot be seen without a telescope. On the other hand, Rigel, which is about 125 times as far away as Proxima Centauri, is so luminous that it is one of the brightest stars in the sky.

Once the distance of a star is known, its real luminosity can be calculated from its apparent brightness at that distance. It turns out that Rigel is about 23,000 times as luminous as our Sun, while our Sun is, in turn, almost 20,000 times as luminous as Proxima Centauri.

All true stars obtain their energy from hydrogen fusion at their centers. Such stars continue shining in a rather steady manner as long as the quantity of hydrogen in their cores remains above a certain amount. During this time, they are said to be on the *main sequence*.

As it happens, the more luminous a star, the more massive it is. (Eddington worked this out when he was calculating the temperature at the center of the Sun.) This means that the more luminous a star is, the more hydrogen it must contain.

You might suppose that this means that the more luminous a star and the more hydrogen it has, the longer it can stay on the main sequence. Actually, the reverse is true. The more massive a star, the more intense is its gravity and the more rapidly it must consume its hydrogen to stay hot enough to resist the gravitational urge to collapse. Although the hydrogen content increases as a star grows

# THE BEGINNING

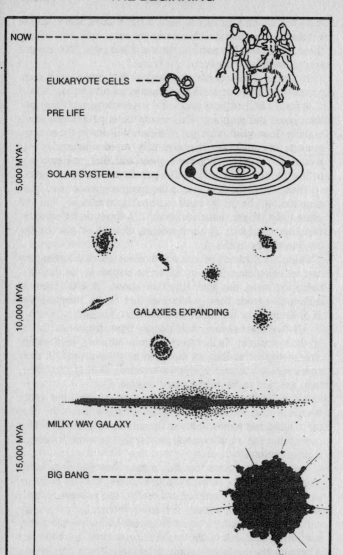

NOW

EUKARYOTE CELLS ---

PRE LIFE

5,000 MYA*

SOLAR SYSTEM ------

10,000 MYA

GALAXIES EXPANDING

MILKY WAY GALAXY

15,000 MYA

BIG BANG - - - - - - - - -

*Millions of Years Ago

larger, more luminous, and hotter, the rate at which hydrogen must be consumed increases much more quickly.

This means that the more luminous a star, the *shorter* its stay on the main sequence.

Our Sun is at a level of luminosity that would keep it on the main sequence for 10,000 million years altogether. It is not quite 5,000 million years old right now, so it is a middle-aged star with a future as long as its past. Once the 10,000 million years are over, the Sun will leave the main sequence and undergo comparatively rapid changes, expanding into a huge, cool *red giant* and then collapsing into a tiny, hot *white dwarf*. Life on Earth will no longer be possible after the Sun leaves the main sequence, but, as I say, that will be 5,000 million years from now.

Sirius, the brightest star in the sky, is about twenty-three times as luminous as the Sun and its lifetime on the main sequence is only 500 million years. It could, at longest reckoning, only have become a star 500 million years ago, when trilobites and ostracoderms swarmed in the early Ordovician seas. Of course, it could easily be even younger than that since there is no sign that Sirius is near the end of its stay on the main sequence. (It has a companion star whose existence may complicate these estimates.)

The most luminous stars we observe are 100,000 times as luminous as the Sun, or more. They must expend their enormous content of hydrogen so rapidly that they can't remain on the main sequence for more than 10 million years or so. After 10 million years, they expand to a red giant, then explode and for a few months shine with the light of a billion stars, then collapse into near invisibility as a neutron star, or into actual invisibility as a black hole.

The most luminous stars we see may have formed after the first hominids appeared on Earth, when our Sun had already been shining steadily for over 4,000 million years.

And if stars have been formed so recently, might it not be that stars are forming now? Right now?

Yes, indeed. There are vast clouds of dust and gas among the stars. One of these is the Orion nebula, and in it are stars, dimly seen through the dust, that may have formed very recently indeed. Then, too, the Dutch-American astronomer Bart Jan Bok (1906–1983) pointed out small round black spots in gas clouds, which are now called *Bok globules*. These may be stars in the process of condensation and formation but in which the cores have not yet grown hot enough to start hydrogen fusion going, so that they are not yet shining.

If, then, stars are forming now, and in the recent past and the not-so-recent past, it seems logical to suppose that stars have been forming steadily ever since the Sun was formed.

In that case, what right have we to think that our Sun wasn't born at a time when other stars already existed? Those other stars may have been luminous ones that after the Sun's formation, but many ages ago just the same, have left the main sequence. Or they might have been very dim stars with extended lifetimes that still exist today and will continue to exist long after our Sun has left the main sequence.

Thus, Proxima Centauri is so dim, and expends its hydrogen in so niggardly a fashion, that it may have a total stay on the main sequence of 200,000 million years. That does not necessarily mean that the Universe must be at least 200,000 million years old. After all, Proxima Centauri must have been formed at the same time as its two companion stars, and one of those companions is just as bright as the Sun so that it can't possibly be more than 10,000 million years old. That means Proxima Centauri can't possibly be more than 10,000 million years old, either, and so it still has 95 percent of its lifetime (such as it is) ahead of it.

From our study of individual stars, then, we know that the Universe is at least 4,550 million years old, since that is how old our solar system is. We know that it is probably older, and even much older. How much older it might be, however, we cannot say from our study of stars alone, and we must look elsewhere.

We can begin with a faintly luminous band that encircles the sky, something that is best seen on a clear moonless night well away from humanity's artificial lighting. The Greeks called it *galaxias kyklos* ("milky circle"). The Romans called it *via lactea* ("milky way"), and we call it, in English, the Milky Way.

Some ancient Greek philosophers thought that the Milky Way might be a crowd of very faint stars, too faint to see individually. That was only speculation, but in 1609, when Galileo turned his telescope on the heavens, he found that speculation to be correct. The Milky Way was indeed composed of innumerable faint stars that melted into featureless luminosity to the unaided eye. In fact, wherever Galileo looked he saw stars, hitherto unseen, crowding among the known stars. The new stars he saw were faint ones—too faint to see without a telescope. Ever since then, better telescopes have made it possible to see more and more, fainter and fainter, stars.

In 1784, the German-English astronomer William Herschel (1738–1822) decided to count the number in each of 683 equally small regions evenly spaced across the heavens. He found that the number of stars in a region far away from the Milky Way was relatively low, but the number increased steadily as one approached that luminous band.

He suggested that the Sun was part of a vast conglomeration of stars that had the shape of a lens (or, to use something more familiar today, a hamburger patty). The Sun is embedded in the lens, and if we look at the sky in the direction of the short diameter of the lens, we see relatively few stars. If we look away from that short diameter, our line of sight travels through longer and longer paths within the lens and we would see more and more stars. Finally, if we looked along the long diameter of the lens we would see so many stars that they would fade into a general luminosity. This conglomeration of stars, of which our solar system is part, is called a *galaxy,* from the Greek expression for the Milky Way.

Herschel tried to estimate the dimensions of the galaxy, and the number of stars it contained, but fell far short of the truth. Later astronomers made better estimates that gave larger figures, but even as late as 1906, they still fell far short.

In 1912, however, the American astronomer Henrietta Swan Leavitt (1868–1921) studied certain stars called *Cepheids*. These were *variable stars,* whose light brightened and dimmed in a regular pattern over a fixed period of time. Some Cepheids were brighter than others, either because some were more luminous than others, or because some were closer to us than others, or both. Given two Cepheids, it was usually impossible to tell if the brighter was brighter because it emitted more light or because it was closer to us.

Leavitt, however, was studying the Cepheids in the Small Magellanic Cloud, a grouping of stars that lay far outside the Milky Way. No matter where particular stars were in the cloud, all were very nearly the same distance from us. (This is analogous to the way in which all the people in Chicago, wherever they might be in the city, are about the same distance from New York.)

In the Small Magellanic Cloud, then, if one Cepheid is brighter than another, it's because the first is the more luminous. Distance doesn't enter into it. Leavitt discovered, then, that the brighter the star, the longer the period in which it carried through its dimming and brightening.

This meant that if we observed any Cepheid anywhere, its period would tell us how luminous it was. Knowing its actual luminosity and its apparent brightness in the sky, we could calculate its distance. (It turned out to be by no means as easy as all that, but astronomers worked out methods for doing this.)

Next, we can turn to another puzzle. There are about a hundred *globular clusters* to be seen in the sky. These are crowded conglomerations of stars in a more or less spherical shape, with each cluster containing tens of thousands of stars. These were first accurately described by William Herschel.

Oddly enough, the globular clusters are not evenly spread over the sky, something that was first pointed out by William Herschel's son, the English astronomer John Herschel (1792–1871). Almost all of them were in one hemisphere of the sky, and fully one-third of them were in the single constellation of Sagittarius, which takes up only 2 percent of the sky.

After Leavitt made her discovery about the Cepheids, another American astronomer, Harlow Shapley (1885–1972), used her findings to measure the actual distance of the different globular clusters. He located Cepheids in each one, measured their period of variation and their apparent brightness, then calculated their distance. This enabled him to make a three-dimensional model.

It turned out that the globular clusters were grouped in a large sphere centered about some spot in the galaxy that was about 30,000 light-years away from us in the direction of the constellation Sagittarius. In 1918, Shapley reported that this spot had to be the center of the galaxy. We couldn't see it (let alone anything on the other side of the galaxy, beyond the center) because of dark clouds of dust and gas that lie between ourselves and the center.

Our solar system is located in the outskirts of the galaxy, in other words, well away from the center, and all we see is our own part of the structure. Earlier astronomers had thought the limited portion we could see without interference from dark clouds was all there was, and that was why they kept underestimating the size of the galaxy.

Our galaxy is now believed to be about 100,000 light-years from end to end along the long diameter. At the center of the galaxy, it is about 16,000 light-years thick, but out here in the outskirts where the Sun is, the Galactic lens has thinned down to where it is only 3,000 light-years thick.

The total mass of our galaxy is equal to 100,000 million times that of the Sun. The average star, however, is considerably less massive than the Sun, so that the galaxy may contain 200,000 million stars, or even more.

Outside our galaxy is the Small Magellanic Cloud, which is 165,000 light-years away, and near it is the Large Magellanic Cloud, which is 155,000 light-years away.

They are small galaxies, each containing between 1,000 million and 10,000 million stars.

Is there anything else in the Universe besides our galaxy and the Magellanic Clouds? Harlow Shapley and most astronomers in the 1910s thought there wasn't. The galaxy and the Magellanic Clouds, they thought, comprised the entire Universe.

In opposition was an American astronomer, Heber Doust Curtis (1872–1942). Whereas Shapley and others thought that the Andromeda nebula was a cloud of dust and gas that was part of our galaxy and not very distant, Curtis thought it was a collection of stars, so far off that even the finest telescopes could not make them out as single bits of light.

Curtis's evidence was this. While ordinary stars in the Andromeda nebula were too distant to make out singly, every once in a while a star flares up to unusual bright-ness. We call these stars *novas* (from the Latin word for "new," because in older times such a flaring star might convert one that was ordinarily invisible into one that was quite bright for a time. It would then seem a new star in the sky.)

There are novas in our own galaxy, but they appear only occasionally in various parts of the sky. No one part gets very many. Curtis, however, when he was watching the Andromeda nebula, would see frequent little dots of light that he could just barely make out in his telescope. These, he maintained, were novas. There were so many of them in that one little patch of sky taken up by the Andromeda nebula, and they were so faint, that they couldn't be stars of our own galaxy; they had to be stars of the nebula itself which had to be a far distant galaxy, in that case, one that was far more distant from us than the Magellanic Clouds.

Curtis and Shapley held an important debate on the subject in 1920, and Curtis did surprisingly well, holding his own against Shapley and presenting his evidence in a strong manner. Nevertheless, the matter couldn't be settled simply by a debate.

However, in 1917 a new telescope had been installed on

Mount Wilson, just northeast of Pasadena in California. Its mirror was 100 inches (154 centimeters) across, a world record at the time, and it could make out things more clearly and at greater distances than any other telescope then in existence.

Making use of the telescope was the American astronomer Edwin Powell Hubble (1889–1953). By 1923, he took photographs of the Andromeda nebula, which showed that it consisted of a mass of fantastically dim stars. He identified some of the stars as Cepheids, and once he had measured their period, he could work out their distance. Curtis, it turned out, was right. The Andromeda nebula was an extremely distant collection of stars, closely resembling our own galaxy. It was, in short, another galaxy. It is now called the Andromeda galaxy and our own galaxy is often called the Milky Way galaxy as a way of distinguishing it from other such objects.

Nor was the Andromeda galaxy the only one of the sort. Once it was understood that there were galaxies other than our own, many other nebulas were recognized as distant galaxies, and almost all of them proved to be far more distant even than the Andromeda. There are millions of galaxies. Indeed, it is frequently estimated that there may be as many as 100,000 million galaxies.

It was only in the 1920s, then, that human beings finally began to get a glimpse of the true size of the Universe. Instead of thinking of the Universe as a collection of individual stars, astronomers began to think of it as a collection of galaxies, and even of clusters of galaxies, and that helped them understand some matters much better.

For instance, there is no way of estimating the age of the Universe by studying the stars of the Milky Way galaxy, but it could be done by studying the different galaxies.

The method of doing so dates back to a discovery by the Austrian physicist Christian Johann Doppler (1803–1853). In 1842 he showed that the pitch of sound changed if the source of the sound was moving with respect to the listener. If the source was moving toward the listener, the sound waves were squeezed together and grew shorter, and therefore higher in pitch. If the source was moving away from the listener, the sound waves were stretched and grew longer, and therefore deeper in pitch. This is called the *Doppler effect*. (Naturally, this is best heard when one is dealing with a single wavelength of sound.)

In 1848 a French physicist, Armand Hippolyte Fizeau (1819–1896), pointed out that the Doppler effect ought to work for light as well, and so it does. When a light source is moving away from you, the light waves grow longer and therefore move in the direction of redness since red is what we see when the light waves are particularly long. When a light source moves toward you, the light waves grow shorter and therefore move in the direction of violetness, since violet is what we see when light waves are particularly short.

This would work for stars, but stars send out all sorts of wavelengths of light in a complicated jumble and it is hard to tell any change in that jumble.

However, when the light from a star (or from any source) is passed through an instrument called a *spectroscope,* the light waves are spread out in order, the longest waves of red at one end and the shortest waves of violet at the other end, with the light waves changing smoothly in length from one end to the other. The result is a rainbow of colors—red, orange, yellow, green, blue, and violet—called a *spectrum.*

The spectrum is often missing certain wavelengths that atoms in the light source have absorbed. These missing wavelengths show up as dark lines in the spectrum. These lines were first discovered by a German optician, Joseph von Fraunhofer (1787–1826), in 1814.

Each element produces certain dark lines that no other element produces, and these dark lines are always in the same place, provided the light source isn't moving with respect to the observer. That place can be measured accurately. If the light source is receding, the dark lines move toward the red end of the spectrum and this is called the *red-shift.* If the light source is approaching, the dark lines move toward the violet end of the spectrum and this is the *violet-shift*.

The greater the red-shift the faster the light source is receding, and the greater the violet-shift the faster it is approaching. This works at any distance, provided you can form a spectrum of the distant light source.

This isn't so easy, but astronomers learned to make tiny spectra out of the light of a single star. More important still, after photography was invented in 1839 by the French inventor Louis Jacques Daguerre (1789–1851), astronomers learned how to take photographs of these tiny spectra, study the dark lines in them, and measure the positions in order to see in which direction they had shifted and by how much. In this way, they could tell how quickly a star was receding or approaching.

The first successful use of this technique came in 1868 when the English astronomer William Huggins (1824–1910) measured the shift of the dark lines in the spectrum of the bright star Sirius and found that it was receding.

As the technique improved, spectra of dimmer and dimmer stars were studied. Some were found to be approaching and some receding, some at relatively low speeds and some at speeds of 65 miles (100 kilometers) per second or more.

Then, in 1912 an American astronomer, Vesto Melvin Slipher (1875–1969), studied the spectrum of the Andromeda nebula, which was not yet known to be a galaxy. It was an average spectrum of many, many stars, but he found dark lines and he could measure their position. He found that the Andromeda was approaching at a speed of 125 miles (200 kilometers) per second. This was a little fast, but not too much so, and it didn't strike Slipher as anything out of the way.

By 1917, however, things seemed somewhat more puzzling. Slipher had gone on to measure the motion of fifteen different nebulas that resembled the Andromeda but were fainter (and, therefore, probably farther away). Of these, Andromeda and one other were approaching and the remaining thirteen were all receding. What's more, those that were receding were doing so at velocities that were unusually high, moving at speeds of 400 miles (640 kilometers) per second and more.

Once it was discovered that these nebulas were really distant galaxies, interest in their motions grew more intense. Another American astronomer, Milton La Salle Humason (1891–1972), took up the task. He made photographic exposures of days at a time in order to get the spectra of very faint galaxies, and they all continued to show red-shifts. The galaxies were all receding, and the fainter they were, the faster they were receding. In 1928 Humason found a galaxy that was receding at a speed of 2350 miles (3800 kilometers) per second, and by 1936 he was clocking recessions at 25,000 miles (40,000 kilometers) per second.

Hubble, who had first seen the stars in the Andromeda, was working along with Humason. He did his best to estimate the distance of various galaxies. For those that were close enough, he used Cepheids. For those that were so far away that all the Cepheids were too dim to be seen, Hubble used the very brightest stars they contained, on the assumption that they would be as luminous as the very brightest stars in our own galaxy. If a galaxy was so distant that not even its brightest stars could be seen, he judged distance from the overall brightness of the entire galaxy.

By 1919, he had enough data to feel justified in announcing that the farther a galaxy was, the faster it was receding. If one galaxy was twice as far from us as another was, the first galaxy receded at twice the velocity of the other. This was called "Hubble's law."

But why should this be? The logical conclusion was that the Universe was expanding.

The galaxies exist in clusters, and within the clusters gravity holds all the galaxies in its grip, so that two galaxies in a cluster might be moving slowly toward each other or away from each other. The Andromeda is in the same cluster that the Milky Way is in, which is why the two are approaching each other slowly. With the passage of time the two may start receding from each other.

Different clusters of galaxies, however, *always* recede from each other. It is not that they are receding from *us;* they are receding from each other. If we were standing in any other galaxy, the distant galaxies would still all seem to be moving away.

Actually, such an expanding Universe had been predicted. In 1916 the German-Swiss physicist Albert Einstein (1879–1955) had worked out his *general theory of relativity* in which he described the workings of gravitation, and just about everything else dealing with the large-scale structure of the Universe, in a series of equations.

The Dutch astronomer Willem de Sitter (1872–1934) pointed out in 1917 that Einstein's equations seemed to predict that the Universe was expanding. There was no indication at the time that this was so, so Einstein introduced a special term in his equations to make it possible to solve them in such a way as to show that the Universe was static. When, eventually, it was clear that the Universe *was* expanding, Einstein removed that special term, calling it the greatest scientific mistake of his life.

But if the Universe is expanding, what if we look deeper and deeper into the far past, as though we were running a motion picture film backward?

Helmholtz had done this when he decided the Sun was contracting. He looked into the past and considered the way the Sun would be expanding. In this way, he calculated the age of the Earth by determining the time it would take the Sun to expand till it filled Earth's orbit under reversed-film conditions.

Then, again, when George Darwin realized that the Moon was moving away from Earth, he looked into the past, reversing the film, and calculated the way in which the Moon would be approaching the Earth. In this way, he decided the Moon was originally part of the Earth.

Both Helmholtz and Darwin had come to the wrong conclusions, but that was not the fault of the notion of reversing the film, but of other complications.

What, then, if we reverse the film of the expanding Universe? If we look backward across the millions of years, we would be watching the Universe contract. We would be watching the clusters of galaxies come closer and closer together until, perhaps, they would all merge with each other, so that all the contents of the Universe would come together in one fat lump.

A Belgian astronomer, Georges Edward Lemaitre (1894–1966), reasoned this way even before Hubble had worked out his law. Lemaitre imagined the original situation with all the contents of the Universe in a lump and called that lump the "cosmic egg." He imagined that this cosmic egg was unstable and that it had exploded. The clusters of galaxies were still flying apart as a result of that unimaginably huge explosion.

The Russian-American physicist George Gamow (1904–1968) was one of the astronomers who grew immediately interested in Lemaitre's suggestion. He called the initial explosion the "Big Bang" and the expression caught on.

Not everyone accepted the Big Bang, of course. It seemed entirely speculative and there was no evidence for it except the fact that the Universe was expanding, and after all perhaps it was just pulsating. It happened to be expanding now for a while, but later it might be contracting for a while, and so on.

Gamow, however, pointed out in 1948 that the Big Bang ought to involve enormous temperatures and radiation and that this should gradually cool down as the Universe expanded. Even today that radiation should exist as a form of radio wave coming equally from all parts of the sky.

In 1964 two American physicists, Arno Allan Penzias (b. 1933 in Germany) and Robert Woodrow Wilson (b. 1936), actually detected this radiation from all parts of the sky. It was almost exactly as Gamow described it. Since then, the notion of the Big Bang has been accepted by just about all scientists.

Theoretical physicists, especially, have tried to work out what conditions must have been like after the Big Bang, and we'll get to that in a little while.

Meanwhile, let's ask the question that anyone interested in beginnings must ask. When did the Big Bang take place?

This can be calculated if one knows the distances between the clusters of galaxies and how quickly they are moving apart from each other. The farther apart they are, the longer it will take them to come together if the film is reversed. The more slowly they are separating, the more slowly they will be coming together if you reverse the film and the longer it will take them to do so.

Hubble had judged the distance of the Andromeda galaxy by the periods and brightnesses of the Cepheids he could detect within it. He ended up with an estimate of 800,000 light-years as the distance of the Andromeda galaxy. This is an enormous distance, five times the distance of the Magellanic Clouds. Other galactic distances were based to some extent on this figure for the Andromeda galaxy.

Using those distances, and the manner in which the speed of recession increased with those distances, the estimate was that, with the film in reverse, all the galaxies would come together in 2,000 million years. That meant that the Big Bang took place, and the Universe began, 2,000 million years ago.

This created the same kind of furor that had taken place eighty years before when the supposedly shrinking Sun made it seem that Earth was no more than 100 million years old. Geologists and biologists knew then that Earth and life were older than 100 million years, and in the 1930s they knew that Earth and life were older than 2,000 million years.

The astronomers held firm on the galactic data for a while, but the matter seemed to be shaky for them in some respects. The Andromeda galaxy was smaller than the Milky Way galaxy, as were all the other galaxies. It seemed rather suspicious that our own galaxy should be so much the largest. Then, too, the Andromeda galaxy has globular clusters just as the Milky Way galaxy does, but the Andromeda's globular clusters seemed far dimmer than our own are.

Could it be that the Andromeda galaxy and all the other galaxies are farther away than was thought? If they were farther away, they would have to be larger to seem the size we perceive them to be, and the globular clusters would have to be more luminous to have the brightness that is observed.

In 1952 the German-American astronomer Walter Baade (1893–1960) studied the Cepheids very carefully and found that there were two varieties. You could calculate distance from one variety according to the equations worked out by Leavitt and by Shapley, but the other variety required a different equation.

Shapley had happened to use the correct variety of Cepheids to work out the size of the Milky Way galaxy and the distance of the Magellanic clouds. Without knowing it, however, Hubble had applied the equations to the other variety of Cepheids in working out the distances of the galaxies. If the new and proper equations were applied to the Cepheids in the Andromeda galaxy, it turned out that it was much farther away than Hubble had thought. Instead of being 800,000 light-years away, it was about 2,300,000 light-years away, about three times farther than had been thought.

In addition, continued investigations of red-shifts, and more refined measurements, make it look as though the galaxies are separating considerably more slowly than Hubble had thought.

Both changes make the time of the Big Bang longer ago than had been thought. Astronomers still don't entirely agree on the time, except that it's long enough to satisfy the geologists and biologists. Some astronomers think the Big Bang took place about 10,000 million years ago, and others think the figure should be 20,000 million years ago. It is safest, perhaps, pending further discoveries, to suppose that it took place 15,000 million years ago.

The Big Bang does leave some problems, though. Astronomers assume that the Universe in its very early days had a smooth and even distribution of matter and energy. Why, then, should the Universe be "lumpy" now, with galaxies and clusters of galaxies separated by empty space?

Then, too, astronomers are not quite sure how much matter and energy there is all told, and just what the average density of matter in the Universe might be. If there is more than a certain amount, the Universe's expansion will very gradually slow until it comes to a halt, and thereafter it will begin to contract again. If there is less than that certain amount, the Universe will expand forever. Apparently the actual density is so close to that certain amount that astronomers can't be sure which alternative is correct. It seems a puzzling coincidence that the density figure should be so close to that certain amount.

Astronomers and physicists have tried to work back toward the Big Bang, assuming that the laws of nature hold no matter how far back they go. They made calculations that dealt with a Universe that grew smaller and smaller as they went farther back in time, and hotter and hotter.

By 1979 they had decided that everything depended on the events in the first second after the Big Bang.

In 1980 the American physicist Alan H. Guth suggested that immediately after the Big Bang there was a period of sudden and vast inflation. In fact, that inflation took place and was *finished* by the time a millionth of a trillionth of a trillionth of a second had passed after the Big Bang. The Universe was then at a temperature of over a trillion trillion degrees. The inflation carried the Universe from a size that was far smaller than a proton to the point where it was 1 centimeter across and from then on it expanded as earlier pictures of the Big Bang had described.

This *inflationary Universe* solved some of the problems the notion of the Big Bang had introduced, but astronomers are still tinkering with it in order to make it more satisfactory still.

But is the Big Bang the true beginning of everything? The Universe might have started as a tiny object with all its enormous mass and energy packed into it, but where did that object come from?

In 1973 the American physicist Edward P. Tryon tackled the problem with the use of *quantum mechanics*. Quantum mechanics is a way of treating the behavior of subatomic particles according to mathematical equations worked out in the 1920s by such scientists as the Austrian physicist Erwin Schrodinger (1887–1961) and the German physicist Werner Karl Heisenberg (1901–1976). Since then, quantum mechanics has proved phenomenally successful and has met every test.

Tryon showed that according to quantum mechanics, it was possible for a Universe to appear, as a tiny object, out of *nothing*. Ordinarily, such a Universe would quickly disappear again, but there were circumstances under which it might not.

In 1982 Alexander Vilenkin combined Tryon's notion with the inflationary Universe and showed that the Universe, after it appeared, would inflate, gaining enormous energies at the expense of the original gravitational field, and would not disappear. However, it would eventually slow its expansion, come to a halt, begin to contract and return to its original tiny size and enormous temperature, and then, in a ''Big Crunch,'' disappear into the nothingness from which it came.

Of course, somewhere in the infinite sea of Nothingness (which somehow reminds one of the infinite sea of Chaos that the Greeks imagined as a starting point) there may be an infinite number of Universes of all sizes beginning and ending—some having done so unimaginably long before our own, and some that will do so unimaginably long after our own.

It does not seem likely, however, that we'll ever know of any other Universes. We may be doomed to know only our own, and we have traced it back to what may well be its absolute beginning some 15,000 million years ago, together with a forecast of what may well be an absolute ending at some undetermined time in the future.

And with that, the business of this book is done.

# INDEX

Abel, 30
Abraham, 22
Acorn worm, 134
Acritarchs, 189
Actinopterygii, 126
Adam and Eve, 28, 143
Aegyptopithecus, 74
Aepyornis, 91
Agamemnon, 32
Agnaths, 131
Agriculture, 27ff
Air, 203ff
Airplane, 1ff
Air pressure, 204
Airship, 3
Airy, George Biddell, 149
Akerblad, Johan David, 17
Albatross, 90
Albuminous substances, 220
Alfven, Hannes, 252
Algae, 124, 188

Alpha Centauri, 257
Alvarez, Walter, 107–08
American Indians, 39
Amino acids, 221
Ammonia, 215
Amphibia, 120ff
    age of, 120
    ancestry of, 128
Amphioxus, 133
Anaerobic bacteria, 237
Anapsida, 97–98, 107
Andrews, Roy Chapman, 79
Andromeda galaxy, 264, 265, 267ff
Andromeda nebula, 249
Angular momentum, 248, 250, 252
Animals, gliding, 89
Ankylosaurus, 103
Annelids, 179–80

Annelid superphylum, 180
Antarctica, 155
Anthropoid apes, 71
Antiquarians, 31
Ape-man, 57
Apes, 68ff
Apollo, 30
Arachnids, 122
Aragon, 8
Archeology, 30
Archeopteryx, 92–93
Archosauria, 98, 103, 104
Arginine phosphate, 178–79
Argon, 211
Argon-40, 164
Aristotle, 70
Arrhenius, Svante August,
    231
Arthropoda, 94, 179
    invasion of land by, 121
Ascension Island, 152
Assyrians, 18
Asteroids, 249
Athena, 29
Atlantis, 140, 141
Atlas, 203
Atmosphere, 203ff
    early, 213ff
    gravity and, 205,
        206
    loss of, 206
    uniqueness of, 211–12
Atoms, 205
Augustus, 11
Aurochs, 77
Australia, 84, 85
Australian aborigines, 39
Australopithecines, 63ff
Avery, Oswald Theodore,
    226
Azores, 152

Baade, Walter, 272
Babylon, 15, 137, 138
Bacon, Francis, 141
Bacteria, 191ff
    anaerobic, 237
Balanoglossus, 134
Balloon, 4
Baluchistan, 79
Baluchitherium, 79
Barbary ape, 70–71
Barghoorn, Elso Sterrenberg,
    189, 194
Bats, 87, 88
Bawden, Frederick Charles,
    198
Becquerel, Antoine Henri,
    162
Bee hummingbird, 90
Behistun, 18
Bijerink, Martinus Willem,
    197
Beijing, 60
Benda, C., 187
Beneden, Edouard van, 186,
    187
Bessel, Friedrich Wilhelm,
    256
Beta Pictoris, 254
Bible, 20ff, 42ff
Big Bang, 270ff
Big Crunch, 274
Bilateral symmetry, 177
Birds, 89ff, 104
    flightless, 90, 92
Black, Davidson, 60
Black, Joseph, 210
Black hole, 259
Blue-green algae, 193
Blue whale, 76–77
Bok, Bart Jan, 260
Bok globules, 260

Boltwood, Bertram Borden, 163
Bonaparte, Napoleon, 16
Bonnet, Charles, 49
Bony fish, 129
Bouchard, 16
Bouguer, Pierre, 149
Boule, Marcellin, 54
Boyle, Robert, 204
Brachiosaurus, 102
Braconnot, Henri, 221
Brontosaurus, 102
Bronze Age, 37, 38
Broom, Robert, 63
Brown, Robert, 185
Bruno, Giordano, 142
Buchner, Eduard, 222
Buffon, Georges Louis de, 143–44, 241, 247–48
Burnet, Thomas, 143

Cable, Atlantic, 151
Caesar, Julius, 12
Cain, 30
Cairns-Smith, A. G., 235
Calcium carbonate, 210
Calcium oxide, 210
Cambrian, 110, 167ff
Canaan, 21, 22
Capillaries, 173
Carbon-14, 35, 36
Carbonaceous chondrites, 234
Carbon dioxide, 210ff
    atmospheric formation and, 217
Carboniferous, 110, 111
Carnosaurs, 99
Carthage, 14
Cartilage, 129, 130, 131

Caspersson, Torbjern Oskar, 199
Castile, 8
Catalysts, 222
Catastrophism, 49, 109
Cave bear, 77
Cayley, George, 4–5
Cell division, 186
Cell membranes, 174
Cells, 173ff
    formation of, 236
Cell theory, 174
Cellulose, 220
Cell walls, 173
Cenozoic, 72
Cepheids, 262, 263, 271
Chaldeans, 18
Chamberlin, Thomas Chrowder, 250, 251
Champollion, Jean Francois, 17
Chaos, 137, 138, 139, 202
Chargaff, Erwin, 226, 227
Charlemagne, 9
Chemical evolution, 235
Chemosynthetic bacteria, 238
Chimpanzee, 71
China, 19, 60
Chinese giant salamander, 121
Chitin, 121
Chlorophyll, 123, 174
    bacteria and, 193
Chloroplasts, 174
Chondrichthyes, 130
Chordata, 94, 132ff
    larvae of, 177
    origin of, 134, 177, 178
    sea, 126ff
Choukoutien, 60

Christian era, 12
Chromatin, 186
    bacteria and, 192
Chromosomes, 186
    heredity and, 199–200,
        224ff
Chronology, 9ff
Cilia, 188
City, 28
Civilization, 28–29
Clay, 235
Cliffs of Dover, 83
Cloning, 175
Coelacanths, 128, 129
Coelenterates, 181
Coenzyme, 224
Cohn, Ferdinand Julius, 191
Collagen, 222
Columbus, Christopher, 6,
    7, 141
Comets, 108
Common era, 12
Compsognathus, 99
Conservation of energy, 160
Consul, 73
Constantinople, 7
Continental drift, 150, 151
Continents, 136ff
Copernicus, Nicolas, 246
Copts, 17
Cosmic egg, 270
Cosmos, 138
Cotylosaurs, 112
Craters, 253
Creatine phosphate, 178,
    179
Creation, 23
Cretaceous, 83
Crete, 32, 141
Crick, Francis H. C., 227
Crocodilia, 104

Cro-Magnon man, 40, 52, 54
Crossopterygians, 128, 129
Cuneiform, 15, 16
Curie, Marie Sklodowska,
    162
Curie, Pierre, 162
Curtis, Heber Doiust, 264,
    265
Cyanobacteria, 193, 237
Cyrenius, 11
Cytoplasm, 187
    RNA in, 199

Daguerre, Louis Jacques,
    267
Dana, James Dwight, 147,
    148
Darius I, 18
D'Arlande, Marquis, 5
Dart, Raymond Arthur, 63
Darwin, Charles Robert, 46,
    51, 146, 147, 159–60,
    241
Darwin, George Howard,
    241, 250, 269
Dating, 33ff
David, 21
Dawson, Charles, 67
Deeps, ocean, 154
de Geer, Gerard, 34
Deinosuchus, 104
de Lamarck, Jean de Monet,
    146
de Laplace, Pierre Simon,
    248–49
Demeter, 30
Deoxyribonucleic acid
    (DNA), 199
Descartes, René, 143
*Descent of Man, The*, 51

Devonian, 110
Devonshire, 110
de Vries, Hugo Marie, 199
Diapsida, 98
Diatase, 222
Dietz, Robert Sinclair, 154
Dinosaurs, 76, 77, 98
    extinction of, 107, 108
Dionysian era, 12
Dionysius Exiguus, 11
Diplodocus, 102
Dirigible, 2, 3
Disease, infectious, 191
Disney, Walt, 103
Disraeli, Benjamin, 51
DNA, 199
    formation of, 230
    heredity and, 226
    replication of, 226, 227
    structure of, 227
*Dr. Jekyll and Mr. Hyde*, 54
Doppler, Christian Johann,
    265
Doppler effect, 266
Double helix, 227
Douglass, Andrew Ellicott,
    34
Dragonfly, 122
Dryopithecus, 74
Dubois, Marie Eugene, 57ff
Duckbill platypus, 85
Dutch East Indies, 58
Dutton, Clarence Edward,
    149
Dyes, synthetic, 185, 186

Earth, age of, 158ff, 164ff
    circumference of, 240
    origin of, 143–44, 157ff
    rotation of, 240–41

Echidna, 85
Echinoderms, 177ff
Echinoderm superphylum,
    179
Echolocation, 88
Ectoderm, 180
Eddington, Arthur Stanley,
    251, 257
Egg cell, 175
Egg, reptilian, 112
Egypt, 15ff
Einstein, Albert, 269
Elasmosaurus, 105
Elford, William Joseph, 197
Elizabeth II, 8
Endoderm, 180
England, 8
Enzymes, 222, 223, 224
Eoanthropus dawsoni, 67
Eocene, 72
Eogyrinus, 121
Eosuchians, 111
Epoch, 72
Era, 72
Eratosthenes, 240
Esarhaddon, 18
Escape velocity, 205
Estivation, 128
Eukaryotes, 183ff
    oldest, 189
    prokaryotes and, 193,
    194
Euryapsida, 98, 105
Eutheria, 83
Everest, George, 149
Evolution, biological,
    45ff
Evolutionary radiation, 82
Ewing, William Maurice,
    152
Exodus, 22

*Face of the Earth, The*, 148
*Fantasia*, 103
Feathers, 89
Feulgen, Robert Joachim, 199
Ferdinand, 8
Ferments, 222
Fertilized ovum, 175
Filtrable virus, 197
Fire, 61
Fish, 126ff
    age of, 120, 126
Fizeau, Armand Hippolyte, 266
Flagellae, 188
Flemming, Walther, 186
Flight, animal, 87ff
Flight, human, 1ff, 88
Flood, Noah's, 23, 140ff, 149
Fluorine, 67
Flying fish, 89
Food, 219ff
Fossils, 47ff
    oldest, 159, 167ff
Fox, Sidney Walter, 236
France, 8
Franklin, Benjamin, 144
Franklin, Rosalind Elsie, 227
Fraunhofer, Joseph von, 266

Galaxies, clusters of, 268–69
    recession of, 268ff
Galaxy, 261
    size of, 263
Galen, 70
Galileo, 142, 173, 240, 247, 261

Gamow, George, 270
Ganymede, 206–07
Gas, 203–04
Gauls, 13
Gegenbaur, Karl, 175
Gelatin, 219–20
Genes, 200
Genesis, 138
Genetic code, 228
Germany, 9
Germ layers, 180, 181
Germ theory of disease, 196
Giants, 66
Gibbon, 57–58
Gibraltar, 70
Giffard, Henri, 3, 4
Gigantopithecus, 66, 67
Gills, 115–16
Gill slits, 133
Glider, 4, 5
Globular clusters, 262
Glucose, 220
Goliath beetle, 95
Gondwanaland, 148, 155
Gorilla, 64, 71, 72
Great apes, 71
Great dying, 79
Great Global Rift, 153
Great white shark, 130
Greeks, 14ff
Greenland, 151
Gregorian calendar, x
Griffith, Fred, 226
Ground sloth, 77
Gurich, Georg Julius Erns, 182
Gutenberg, Johannes, 7
Guth, Alan H., 273

Haeckel, Ernst Heinrich, 57
Half-life, 163, 164

Halley, Edmond, 157, 158, 256
Hammurabi, 19
Hartmann, William K., 243
Hastings, Battle of, 8
Heezen, Bruce Charles, 153
Heisenberg, Werner Karl, 274
Helium, 214
Helmont, Jan Baptista van, 203–04
Hemichordata, 178
Hemoglobin, 221
Henderson, Thomas, 256
Herculaneum, 31
Hercules, 203
Herding, 28ff
Heredity, 200
Herod, 11
Herodotus, 20
Herschel, John, 262–63
Herschel, William, 261, 262
Hesperornis, 92
Hess, Harry Hammond, 154
Hieroglyphics, 15
Himalayan mountains, 155
*Hindenburg*, 3
Hipparchus, 240
Hissarlik, 31
Hitties, 32
Holocene, 72
Holy Roman Empire, 9
Homer, 14
Hominid, 56
Hominoid, 68
  ancestral, 74
Homo erectus, 61
Homo habilis, 62, 63
Homo neanderthalensis, 53
Homo sapiens, 41

Homo sapiens neanderthalensis, 55
Homo sapiens sapiens, 55
Homo troglodytes, 72
Hooke, Robert, 173
Horseshoe crab, 167
Hoyle, Fred, 92
Hubble, Edwin Powell, 265, 268, 270, 271
Huggins, William, 267
Hugh Capet, 8
Humason, Milton La Salle, 268
Humboldt, Friedrich Wilhelm, 146
Hunters and gatherers, 26, 27
Hutton, James, 144, 145
Huxley, Thomas Henry, 53
Hydrogen, 214
  in universe, 214
Hydrogen cyanide, 233
Hydrogen sulfide, 214
Hydrophobia, 196

Ichthyornis, 91
Ichthyosaurs, 106
*Iliad*, 15, 31
Incas, 19
India, 148
Inflationary universe, 273
Ingenhousz, Jan, 213
Insects, 95, 122
Interstellar dust clouds, 234
Invertebrates, 94
Iridium, 108
Irish elk, 77
Iron age, 37, 38
Isabella, 8
Isoprene, 220

Isostasy, 150
Ivanovsky, Dmitri Iosifovich, 196, 197

Japan, 60
Java, 58
Java man, 61
Jawed fish, 131
*Jaws*, 130
Jeans, James Hopwood, 251
Jeffreys, Harold, 251
Jericho, 36
Jesus, 11, 12, 13
Jewish mundane era, 23
Jews, 138
Johannsen, Wilhelm Ludwig, 200
Johnson, Donald, 64
Joshua, 22
Judges, Israelite, 21, 22
Jupiter, 216
    angular momentum of, 250
Jura mountains, 83
Jurassic, 83

Kamen, Martin David, 35
Kant, Immanuel, 248
Kelvin, William Thomson, Lord, 161
Kepler, Johann, 247
Keratin, 222
Kinetic energy, 248
Kinetic theory of heat, 205
King Kong, 66
Kodiak bear, 99
Koenigswald, Gustav von, 59, 65, 66
Komodo, 105

Komodo dragon, 104–05
Kuhne, Wilhelm, 222

Lamprey, 132
Land, life invades, 116ff
Land bridges, 148
Langevin, Paul, 88, 152
Langley, Samuel Pierpont, 2
Large Magellanic cloud, 263
Lartet, Edouard, 52
Larvae, 134, 176–77
Latimer, Miss, 129
Latimeria, 129
Laurasia, 154
Lavoisier, Antoine Laurent, 204, 211
Lead, 163
Leakey, Louis S. B., 62, 73
Leakey, Mary, 73
Leavitt, Henrietta Swan, 262, 263
Lemaitre, Georges Edouard, 270
Lemuria, 148
Lemurs, 74–75, 89
Leo III, 9
Leonardo da Vinci, 2, 48
Lepidosauria, 98, 104
Lewis, G. Edward, 73
Libby, Willard Frank, 36
Life, 213
    classification of, 43ff
    origin of, 219ff
Light year, 256
Lilienthal, Otto, 5
Limbs, vertebrate, 131
Lines, spectral, 266
Linnaeus, Carolus, 44–45
Lizards, 89
Loch Ness monster, 105

Louis Philippe I, 9
Lubbock, John, 38
Lucy, 64, 65
Lungfish, 127
Lyell, Charles, 146

Madagascar, 74, 91, 148
Magendie, Francois, 219
Malpighi, Marcello, 173
Mammals, 76ff
    age of, 79
    flying, 87, 88
    reptilian ancestry of, 106
Mammoths, 52, 77
March, Frederic, 53
Marduk, 137, 138
Margolis, Lynn, 194
Marius, Simon, 249
Mars, atmosphere of, 206,
    218
Mass extinctions, 78, 107ff
Mastodon, 77
Maury, Matthew Fontaine,
    151–52
Maxwell, James Clerk, 205,
    249
Mayans, 19
Mendel, Gregor Johann,
    199, 200
Mercury, 206
Mesoderm, 180–81
Mesozoic, 72
Messenger-RNA, 228
Meteor (ship), 152
Meteorites, 234
Methane, 215
Microorganisms, 188
Microscope, 173
Microspheres, 236
Mid-Atlantic ridge, 152

Middle Ages, 8
Mid-Oceanic ridge, 153
Miescher, Johann Friedrich,
    198
Milky Way, 261
Miller, Stanley Lloyd, 233
Miocene, 72
Missing link, 51
Mitochondria, 187
Mitosis, 186
Moa, 91
Modern man, 36
Molecules, 205, 206
Monkeys, 68ff
Monotremes, 85
Montgolfier, Jacques
    Etienne, 5
Montgolfier, Joseph Michel, 5
Moon
    astronauts on, 243
    distance of, 240
    origin of, 239ff
    size of, 240
Moulton, Forest Ray, 250
Mountains
    density of, 149
    formation of, 147, 151, 155
Mulder, Gerardus Johannes,
    220
Muller, Otto Friedrich, 191
Multicellular organisms,
    172ff
    pre-Cambrian, 182, 183
Mutations, 227
Mycenae, 31

Narmer, 18
Natural selection, 47
Neanderthal man, 52ff
Nebuchadnezzar, 18

Nebular hypothesis, 249
Neolithic, 38
Neoteny, 178
Neptune, 216
Nerve cord, 132
Neutron star, 259
New Guinea, 40, 85
New Stone Age, 38
Newton, Isaac, 205, 240, 248
New Zealand, 40
Nitrogen, 211, 217
Notochord, 94, 133, 134
Novas, 264
*Novum Organum*, 141
Nuclear energy, 163
Nucleic acids, 198ff
    heredity and, 224ff
Nuclein, 198
Nucleoprotein, 199
Nucleotides, 225
    formation of, 233ff
Nucleus, cell, 185
    DNA in, 199

Ocean
    bottom of, 151, 152
    formation of, 203ff
    salt in, 157, 158
    uniqueness of, 208
Old Stone Age, 38
Old Testament, 20ff
Olduvai Gorge, 62
Oligocene, 72
Olympiads, 15
Olympian games, 14
Oparin, Alexander Ivanovich, 233
Orangutan, 71
Ordovician, 110

*Origin of Life on Earth, The*, 233
*Origin of Species, The*, 46, 47, 50
Orion nebula, 259
Ornithischia, 98
Oro, Juan, 233
Osteichthyes, 129
Ostracoderms, 131, 132
Ostrich, 91
Oxygen, 115, 116, 211, 212, 213
    atmosphere and, 217, 237, 238

Pakistan, 79
Palade, George Emil, 187
Paleocene, 72
Paleolithic, 38
Paleozoic, 72, 110
Palissy, Bernard, 142, 143
Pangaea, 151
    breakup of, 155
Panthalassa, 151
Panthotheria, 85
Parallax, 256
Parapsida, 98, 106
Parker, Eugene Newman, 215
Pasteur, Louis, 191, 196, 231
Payen, Anselme, 222
Pegasus, 88
Peking, 60
Peking man, 60
Penguins, 92
Penzias, Arno Allan, 270
Pepsin, 222
Peripatus, 179–80
Permian, 110, 111

Phalangers, 89
Phoenicians, 15
Photography, 267
Photolysis, 217
Photosynthesis, 213, 237
Phylum, 94
Pilatre de Rozier, Jean
    Francois, 5
Piltdown man, 67ff
Pineal gland, 105
Pirie, Norman Wingate,
    198, 199
Pithecanthropus, 57
Pithecanthropus erectus, 59
Placenta, 83
Placental mammals, 83–84
Placoderms, 129, 130
Planetesimal hypothesis, 251
Planetesimals, 251ff
Planets, 246
    orbits of, 247
Plants, 123ff
    invasion of land by, 123,
        124–25
    oxygen and, 213
Plato, 140, 141
Playfair, John, 145
Pleistocene, 72, 73
Plesiosaurs, 105, 106
Pliocene, 72, 73
Pliopithecus, 74
Pneumonia bacteria, 226
Pocahontas, 13
Polymers, 220, 221
Pompeii, 31
Pongid, 73
Ponnamperuma, Cyril, 234
Porifera, 181
Portugal, 8
Poseidon, 30
Potassium-40, 164

Pre-Cambrian, 170
Printing, 7
Proconsul, 73–74
Priestley, Joseph, 211, 212
Primates, 68
*Principles of Geology, The*,
    146
Prokaryotes, 190ff, 192,
    236–37
Prometheus, 29
Prosthetic group, 224
Proteinoids, 236
Proteins, 220
    enzymes and, 223
Protestant Reformation, 7
Protoplasm, 184, 185
Protozoa, 188
Prout, William, 220
Proxima Centauri, 257, 260
Pteranodon, 93
Pterosaurs, 93, 94, 104
Ptolemy V, 16
Purgatorius, 75
Purkinje, Jan Evangelista,
    184
Pyramids, 18, 30

Quantum mechanics, 274
Quirinius, 11

Rabies, 196
Radial symmetry, 177
Radioactive series, 163
Radioactivity, 162
Ramapithecus, 73
Rameses II, 22
Ramsay, William, 211
Rat, 78
Ratites, 90, 91

Rawlinson, Henry Creswicke, 18
Ray, John, 48
Rayleigh, John William Strutt, Lord, 211
Red giant, 259
Red kangaroo, 84
Red Sea, 155
Red shift, 266–67
Relativity, 269
Replication, 226
Reptiles, 97ff
  age of, 82
  early, 111ff
Rhipidistians, 128
Rhynchocephalia, 105
Ribonucleic acid, *see* RNA
Ribosomes, 187
  bacteria and, 193
Ricketts, Howard Taylor, 201
Rickettsia, 201
Rigel, 257
RNA, 199
  heredity and, 228ff
Roc, 91
Rocks, age of, 165
Roman Empire, 10
Roman era, 11
Roman republic, 13
Rome, 10, 11
Rose, William Cumming, 221
Rosetta stone, 16ff
Rubber, 220
Rubidium-87, 164
Rutherford, Daniel, 210
Rutherford, Ernest, 162

Sarcopterygii, 127
Sargon, 19

Satan, 48
Saturn, 216
  rings of, 249
Saul, 21
Saurischia, 98
Sauropoda, 98
Schleiden, Matthias Jakob, 173, 174
Schliemann, Heinrich, 31
Schrodinger, Erwin, 274
Schulze, Max J. S., 184
Schwann, Theodor, 174, 222
Sclater, Philip Lutley, 148
Scorpions, 122
Sea, life in, 114ff
Sea-floor spreading, 154
Sea-urchins, 177, 179
Seaweed, 124
Sedimentary rock, 50
Seeds, 187, 188
Segmentation, 179
Shapley, Harlow, 263ff, 272
Sharks, 130, 131
Shrews, 77
Siberia, 40
Siebold, Karl Theodor Ernst von, 189
Silurian, 110
Sinanthropus pekinensis, 60
Sinbad, 91
Sinsheimer, Robert Louis, 229
Sirius, 256
  luminosity of, 259
  recession of, 267
Sitter, Willem de, 269
Sivapithecus, 73
61 Cygni, 256, 257
Skeleton, internal, 132
Slipher, Vesto Melvin, 267

Small Magellanic Cloud, 262, 263
Smith, J.L.B., 128
Smith, John, 13
Smith, William, 49–50
Snider-Pellegrini, Antonio, 149
Socrates, 18
Solar system, 246ff
Solar wind, 215
Solomon, 70
Solon, 15
Sonar, 88, 152
South Africa, 63
South America, 40
Spain, 8
Spectroscope, 266
Spectrum, 266
Sperm cell, 175
Sphenodon, 105
Spiders, 122
Spitzer Lyman, Jr., 251
Sponges, 181, 182
Spontaneous generation, 231
Sprigg, R. C., 182
Springtails, 122
Squamata, 104
Standing upright, 64, 65
Stanley, Wendell Meredith, 198
Starch, 220
Starfish, 177, 178
Stars
　distance of, 256, 257
　formation of, 259–60
　lifetime of, 259
　luminosity of, 257, 259
　motion of, 256
　variable, 262
Stegosaurus, 102–03
Steno, Nicolaus, 143

Stone Age, 37, 38
Strata, 49, 50
Stromatolites, 194–95
Strontium-87, 164
Suess, Eduard, 148
Sulfur dioxide, 217
Sumerians, 19, 29, 32
Sumner, James Batcheller, 223
Sun, 247
　central temperature of, 251
　chemical makeup of, 215
　energy of, 160, 161, 162
Sunfish, 127
Sutton, Walter Stanborough, 199
Symmetry, 177
Synapsida, 98, 106

Tanzania, 62
Tarquinius Superbus, Lucius, 13
Tartessus, 70
Tasmania, 40
Taylor, Frank Bursley, 150
Tectonic plates, 153
Telegraph Plateau, 152
Telescope, 173
Tennyson, Alfred, Lord, 137
Tethys Sea, 148
Thecodonts, 111
*Theory of the Earth, The*, 145
Thera, 141
Theriodonts, 106
Thermodynamics, 160
Theropoda, 98
Thomsen, Christian Jurgenson, 37
Thorium, 162

Threonine, 221
Thutmose III, 18
Tiamat, 137, 138
Tides, 240, 241
Titan, 207
  atmosphere of, 218
Titanotheres, 78
Tobacco mosaic disease,
  196, 197
Torricelli, Evangelista, 204
Transfer-RNA, 229
Transforming principle, 226
Tree rings, 34, 35
Tree shrews, 75
Triassic, 83
Triceratops, 103
Trilobites, 167
Trinil, 58
Tristan da Cunha, 152
Triton, 207
Trojan war, 15, 32
Troy, 31
Tryon, Edward P., 274
Tuatara, 105
Tunicate, 134
Turtles, 107
Tyrannosaurus Rex, 99

Ultrasonic sound, 88
Ultraviolet light, 217
Uniformitarian principle, 145
Universe
  age of, 271ff
  chemical makeup of, 214,
    215
  expanding, 268ff
  galaxies in, 264, 265
  origin of, 255ff
  planets in, 253, 254
Ur, 33

Uranium, 162
Uranus, 216
Urease, 223
Urey, Harold Clayton, 233
Ussher, James, 21ff

Vacuum, 205
van Helmont, Jan Bapista,
  203
van Leeuwenhoek, Anton,
  188, 191
Varro, Marcus Terentius,
  10–11
Varro, era of, 11
Varves, 33
Vega, 254
Venus, 218
Vertebrae, 94, 132
Vertebrates, 94, 132–33
Vesuvius, Mount, 31
Vilenkin, Alexander,
  274
Violet shift, 267
Virchow, Rudolf, 53
Viruses, 196ff
  crystalline, 198
  size of, 197
Volcanoes, 153–54
von Mohl, Hugo, 184

Wales, 110
Washington, George, 13
Waste, animal, 116
Water
  early atmosphere and, 215
  life and, 207, 209
  on other worlds, 208
  photolysis of, 217
Water vapor, 207

Watson, James Dewey, 227
Wegener, Alfred Lothar, 150, 151
Weidenreich, Fritz, 60
Weizsacker, Carl Friedrich von, 252
Whale shark, 131
White dwarf, 259
William of Normandy, 8
Willstatter, Richard, 223
Wilson, Robert Woodrow, 270
Wodehouse, P. G., 9
Woolley, Charles Leonard, 32
World island, 40

Wright, Orville, 1, 2
Wright, Wilbur, 1, 2
Writing, 19

X-ray diffraction, 227

Yeast, 222
Young, Thomas, 17

Zeppelin, Ferdinand von, 2, 3, 4
Zeus, 203
Zhoukoudian, 60

# INFORMATIVE AND FUN READING

__THE ANIMAL RIGHTS HANDBOOK by Laura Fraser, Joshua
Horwitz, Stephen Tukel and Stephen Zawistowski
0-425-13762-7/$4.50
If you love animals and want the facts about how the fashion, food, and
product-testing industries exploit animals for profit, this book offers step-by-
step guidelines to save animals' lives in simple, everyday ways.

__THE RAINFOREST BOOK by Scott Lewis/Preface by Robert
Redford     0-425-13769-4/$3.99
Look into the spectacular world of tropical rainforests--their amazing
diversity, the threats to their survival, and the ways we can preserve them
for future generations. This easy-to-read handbook is full of practical tips
for turning your concern for rainforests into action.

__MOTHER NATURE'S GREATEST HITS by Bartleby Nash
0-425-13652-3/$4.50
Meet the animal kingdom's weirdest, wackiest, wildest creatures! Learn
about dancing badgers, beer-drinking raccoons, 180-foot worms, Good
Samaritan animals and more!

__FOR KIDS WHO LOVE ANIMALS by Linda Koebner with the
ASPCA   0-425-13632-9/$4.50
Where and how do animals live? How did they evolve? Why are they
endangered? Explore the wonders of the animal kingdom while you
discover how to make the Earth a safer home for all animals.

__SAFE FOOD by Michael F. Jacobson, Ph.D., Lisa Y. Lefferts and
Anne Witte Garland 0-425-13621-3/$4.99
This clear, helpful guide explains how you can avoid hidden hazards--and
shows that eating safely doesn't have to mean hassles, high prices, and
special trips to health food stores.

---